RENEWABLE ENERGY: RESEARCH, DEVELOPMENT AND POLICIES

UNITED STATES FOSSIL FUEL EXPORTS

DEVELOPMENTS, CONSIDERATIONS, AND REGULATIONS

RENEWABLE ENERGY: RESEARCH, DEVELOPMENT AND POLICIES

Additional books in this series can be found on Nova's website under the Series tab.

Additional Ee-books in this series can be found on Nova's website under the e-book tab.

RENEWABLE ENERGY: RESEARCH, DEVELOPMENT AND POLICIES

UNITED STATES FOSSIL FUEL EXPORTS

DEVELOPMENTS, CONSIDERATIONS, AND REGULATIONS

MITCHELL P. HENNING
EDITOR

NOVINKA

New York

For permission to use material from this book please contact us:
Telephone 631-231-7269; Fax 631-231-8175
Web Site: http://www.novapublishers.com

Additional color graphics may be available in the e-book version of this book.

Library of Congress Cataloging-in-Publication Data

ISBN: 978-1-63117-198-7

Published by Nova Science Publishers, Inc. † New York

CONTENTS

PREFACE

Recent technological developments have led to an increase in the domestic supply of natural gas. As a result, there is interest among some parties in exporting liquefied natural gas (LNG) to take advantage of international markets. This has placed new attention on the laws and regulations governing the export of natural gas as well as other fossil fuels. In most cases, export of fossil fuels requires federal authorization of both the act of exporting the fuel and the facility that will be employed to export the fuel. This book discusses federal permitting and oversight of export of fossil fuels; new opportunities and uncertain outcomes of United States natural gas exports; United States coal exports; and coal as a key player in expanded United States energy exports.

Chapter 1 – Recent technological developments have led to an increase in the domestic supply of natural gas. As a result, there is interest among some parties in exporting liquefied natural gas (LNG) to take advantage of international markets. This has placed new attention on the laws and regulations governing the export of natural gas as well as other fossil fuels.

In most cases, export of fossil fuels requires federal authorization of both the act of exporting the fuel and the facility that will be employed to export the fuel. For example, the export of natural gas is permitted by the Department of Energy's Office of Fossil Energy, while the construction and operation of the export facility must be authorized by the Federal Energy Regulatory Commission (FERC). Oil exports are generally forbidden, but an export that falls under one of several exemptions to the ban can be authorized by the Department of Commerce's Bureau of Industry and Security, while oil pipelines that cross international borders must be permitted by the State Department. Coal exports do not require special authorization specific to the commodity; however, as with natural gas and crude oil, other generally applicable federal statutes and regulations may apply to the export of coal.

Restrictions on exports of fossil fuels could potentially have implications under international trade rules. They may possibly be inconsistent with the most favored nation requirement of Article I of the General Agreement on Tariffs and Trade 1994 (GATT 1994) if certain World Trade Organization (WTO) members are treated differently than others. Limits on exports could also potentially violate the prohibition on export restrictions contained in Article XI of the GATT 1994 if they prescribe vague and unspecified criteria for export licensing. However, an export licensing regime does not appear to constitute a "subsidy" to downstream users of fossil fuels under WTO rules.

Article XXI, the exception for essential security interests, may provide justification for potential violations of GATT Articles I and XI. The United States has traditionally considered this exception to be self-judging. However, it is possible that a panel or the Appellate Body might scrutinize the United States' use of the exception.

Article XX of the GATT provides additional exceptions that a member country may invoke if it is found to be in violation of any GATT obligations. For example, WTO Members may maintain an otherwise GATT inconsistent measure if it is necessary to protect an exhaustible natural resource or necessary to protect human health or the environment. Article XIII requires that if an otherwise GATT inconsistent measure is permitted to remain in force due to an Article XX exception, the measure must be administered in a non-discriminatory manner. Export restrictions that treat WTO Members differently would appear not to satisfy the non-discriminatory requirements of Article XIII.

Chapter 2 – As estimates for the amount of U.S. natural gas resources have grown, so have the prospects of rising U.S. natural gas exports. The United States is expected to go from a net importer of natural gas to a net exporter by 2020. Projects to export liquefied natural gas (LNG) by tanker ship have been proposed—cumulatively accounting for about 41.4% of current gross U.S. natural gas production—and are at varying stages of regulatory approval. Projects require federal approval under Section 3 of the Natural Gas Act (15 U.S.C. §717b), with the U.S. Department of Energy's Office of Fossil Energy and the Federal Energy Regulatory Commission being the lead authorizing agencies. Pipeline exports, which accounted for 98% of all exports of U.S. produced natural gas in 2012, are also likely to continue rising. What effect exporting natural gas will have on U.S. prices is a central question in the debate over whether to export. A significant rise in U.S. natural gas exports would likely put upwards pressure on domestic prices, but the magnitude of any rise is currently unclear. There are numerous factors that will affect prices: export volumes, economic growth, differences in local markets, and government regulations, among others. With recent natural gas prices relatively low compared to global prices and historically low for the United States, producers are looking for new markets for their natural gas. Producers contend that increased exports will not raise prices significantly as there is ample supply to meet domestic demand, and there will be the added benefits of increased revenues, trade, and jobs, and less flaring. Consumers of natural gas, who are being helped by the low prices, fear prices will rise if natural gas is exported.

Electric power generation represents potentially the greatest increase in natural gas consumption in the U.S. economy, primarily for environmental reasons. Natural gas emits much less carbon dioxide and other pollutants than coal when combusted. Other types of consumption are not likely to increase natural gas demand domestically for a long time. Use in the transportation sector to displace oil is likely to be small because expensive new infrastructure and technologies would be required. There is discussion of a possible revival of the U.S. petrochemicals sector, but the potential extent of a change is unclear. Getting natural gas to markets where it can be consumed, whether domestically or internationally, may be the industry's biggest challenge. Infrastructure constraints, environmental regulations, and other factors will influence how the market adjusts to balance supply and demand.

Environmental groups are split regarding natural gas use, with some favoring increased use to curb emissions of certain pollutants, while others oppose expanded use of natural gas because it is not as clean as renewable forms of energy, such as wind or solar. The use of hydraulic fracturing to produce shale gas has also raised concerns among environmental groups particularly concerned with its possible impacts on water quality.

The possibility of a significant increase in U.S. natural gas exports will factor into ongoing debates on the economy, energy independence, climate change, and energy security. As the proposed projects continue to develop, policymakers are likely to receive more inquiries about these projects. Proposals to expedite and expand LNG exports have been raised in the 113th Congress, including in S. 192 and H.R. 580. Two other bills, H.R. 1189 and H.R. 1191, would reform the DOE's process for determining the public interest regarding LNG exports and prohibit exports of natural gas produced on federal lands.

Chapter 3 – The gap between available U.S. coal supply and demand may continue to widen as low cost natural gas becomes more attractive to electric power plants and uncertainties with emission regulations may inhibit new coal plant investments. Coal producers with excess supply will likely seek to expand their market abroad. Consequently, U.S. coal exports are forecast to continue to rise over the next decade and possibly longer. Growth potential in Asian markets seems large, but there are potential bottlenecks, such as infrastructure and potential rising costs of regulation, competition from other fuels, and transportation constraints that could slow export growth.

The U.S. Energy Information Administration (EIA), in its 2013 Annual Energy Outlook Early Release reference case, projects net exports of coal to trend up through 2040, almost 50% from 2011 levels, with some fluctuations. The significance of this may have short- and long-term ramifications as well as positive and negative consequences. Increased net exports could improve the U.S. trade balance as well as add government revenue from production that may otherwise decline because of falling domestic consumption. Environmentally, exporting coal may run counter to the current Administration's domestic environmental policies and affect U.S. efforts to address global environmental issues. Depending upon the nature of the coal exports, certain parts of the country may benefit economically. Current and projected coal exports, the associated infrastructure, and the environmental consequences have prompted interest by Congress.

The United States has been exporting coal since the late 1800s. From 2003 to 2012, U.S. coal exports have risen over 200%, mainly driven by competitive production costs, global demand, and lower prices, among other factors. Coal exports comprised 12% of U.S. coal production in 2012. In 2011, U.S. coal exports broke 100 million short tons (MST) for the first time since 1992 and in 2012 surpassed their peak of almost 113 MST in 1981. The value of U.S. coal exports has increased, rising from under $10 billion in 2010 to almost $16 billion in 2011, according to U.S. Energy Information Administration data.

Many factors will influence how much coal will be exported from the United States. Enough projects have been proposed that by 2016 the United States could be exporting more than double its current coal exports. Projects in the Pacific Northwest have attracted much of the attention, even though the Northeast continues to be the source for most exports.

Representatives from state and local agencies, particularly in Washington and Oregon, as well as industry, community, and environmental groups, along the potential coal transport corridor, have expressed outright support for or opposition to port terminal projects that would allow for increased export of

Powder River Basin coal through the Pacific Northwest. Opponents have argued that increased train and barge traffic will have significant adverse impacts to the human, natural, and cultural environment. Project supporters have argued that the projects would create or maintain jobs in the construction, mining, and transportation industries and bring increased tax revenue to the states. Broadly, the National Environmental Policy Act (NEPA) requires federal agencies to consider the environmental impacts of their actions before a final decision is made regarding that action.

Chapter 4 – In a November 2012 report that received significant attention in the media, the International Energy Agency (IEA) predicted that the United States will be nearly energy self-sufficient by the year 2035. Justifications for this claim include reports of declining oil consumption, the unlocking of natural gas resources through new technologies, and overall increases in U.S. energy production. As natural gas becomes more inexpensive and renewable energy sources continue to increase in usage, the energy mix in the United States (and other countries) is expected to change. The IEA estimates that 40 percent of world electricity generation was fueled by coal in 2011. Coal is the only major energy source for which the United States demonstrates a trade surplus. In 2011, the last calendar year for which data are available, the United States exported more than $16 billion worth of coal, while importing only slightly less than $3 billion in coal. Furthermore, coal is expected to remain one of the largest fuel sources in worldwide energy consumption for at least the next two decades. Second only to oil in meeting the energy needs of the world, coal generates more electricity for the United States and the world than any other single fuel. With the key role that coal currently plays in the U.S. trade balance, and is expected to play in the future, it is interesting to look at recent volatile export price movements of this important U.S. resource.

In: United States Fossil Fuel Exports
Editor: Mitchell P. Henning

ISBN: 978-1-63117-198-7

Chapter 1

FEDERAL PERMITTING AND OVERSIGHT OF EXPORT OF FOSSIL FUELS*

Adam Vann, Daniel T. Shedd and Brandon J. Murrill

SUMMARY

Recent technological developments have led to an increase in the domestic supply of natural gas. As a result, there is interest among some parties in exporting liquefied natural gas (LNG) to take advantage of international markets. This has placed new attention on the laws and regulations governing the export of natural gas as well as other fossil fuels.

In most cases, export of fossil fuels requires federal authorization of both the act of exporting the fuel and the facility that will be employed to export the fuel. For example, the export of natural gas is permitted by the Department of Energy's Office of Fossil Energy, while the construction and operation of the export facility must be authorized by the Federal Energy Regulatory Commission (FERC). Oil exports are generally forbidden, but an export that falls under one of several exemptions to the ban can be authorized by the Department of Commerce's Bureau of Industry and Security, while oil pipelines that cross international borders must be permitted by the State Department. Coal exports do not require special authorization specific to the commodity; however, as with natural

* This is an edited, reformatted and augmented version of the Congressional Research Service publication, No. R43231, dated September 17, 2013.

gas and crude oil, other generally applicable federal statutes and regulations may apply to the export of coal.

Restrictions on exports of fossil fuels could potentially have implications under international trade rules. They may possibly be inconsistent with the most favored nation requirement of Article I of the General Agreement on Tariffs and Trade 1994 (GATT 1994) if certain World Trade Organization (WTO) members are treated differently than others. Limits on exports could also potentially violate the prohibition on export restrictions contained in Article XI of the GATT 1994 if they prescribe vague and unspecified criteria for export licensing. However, an export licensing regime does not appear to constitute a "subsidy" to downstream users of fossil fuels under WTO rules.

Article XXI, the exception for essential security interests, may provide justification for potential violations of GATT Articles I and XI. The United States has traditionally considered this exception to be self-judging. However, it is possible that a panel or the Appellate Body might scrutinize the United States' use of the exception.

Article XX of the GATT provides additional exceptions that a member country may invoke if it is found to be in violation of any GATT obligations. For example, WTO Members may maintain an otherwise GATT inconsistent measure if it is necessary to protect an exhaustible natural resource or necessary to protect human health or the environment. Article XIII requires that if an otherwise GATT inconsistent measure is permitted to remain in force due to an Article XX exception, the measure must be administered in a non-discriminatory manner. Export restrictions that treat WTO Members differently would appear not to satisfy the non-discriminatory requirements of Article XIII.

INTRODUCTION

Partly as a result of the increased use of horizontal drilling and hydraulic fracturing to extract natural gas from shale formations in the United States, the domestic supply of natural gas has increased relative to demand, leading to lower domestic prices. This has generated increased interest by some U.S. companies in exporting liquefied natural gas (LNG) to take advantage of relatively higher prices in world markets.[1] This new interest in exporting natural gas has also produced renewed interest in the laws and regulations governing the export of other fossil fuels, including crude oil, natural gas, and coal.[2] This report reviews federal laws and the regulatory regime governing the export of natural gas, crude oil, and coal. This report provides an overview of federal laws and regulations and agency roles in authorizing and regulating the export of these fossil fuels. The report addresses several categories of federal laws and regulations, including (1) statutes that establish the authorization process for the actual export of any of the three listed fossil fuels; (2) statutes that govern the permitting of the facilities that export any of the listed fossil fuels; and (3) generally applicable trade statutes and treaties that affect exports of fossil fuels.

GENERALLY APPLICABLE EXPORT REQUIREMENTS

In general, transactions involving the export of items from the United States to a foreign country are subject to the Export Administration Regulations (EAR) enforced by the Department of Commerce's Bureau of Industry and Security (BIS).[3] However, transactions that fall within the scope of the EAR do not necessarily require an export license from BIS.[4] Whether an export license is required depends on several factors, including the nature of the item, its end use, and its ultimate destination.[5] The EAR provides instructions for exporters to follow when determining whether an export transaction is subject to the EAR and, if so, whether the transaction requires a license.[6] Other general requirements may apply to transactions involving the export of items from the United States. For example, for exports of items subject to the EAR that do not take place electronically or in another intangible form, an exporter is required in certain circumstances to submit a Shipper's Export Declaration (SED) or Automated Export System (AES) Record to BIS and the International Trade Administration in the Department of Commerce's Bureau of the Census.[7] A declaration or record typically contains an identification of the exporter and the commodity being shipped; the date of exportation; and the country of ultimate destination, among other information.[8]

STATUTES GOVERNING AUTHORIZATION TO EXPORT FOSSIL FUELS

Crude Oil

The Energy Policy and Conservation Act of 1975[9] directed the President to "promulgate a rule prohibiting the export of crude oil and natural gas produced in the United States, except that the President may ... exempt from such prohibition such crude oil or natural gas exports which he determines to be consistent with the national interest and the purposes of this chapter."[10] The act further provides that the exemptions to the prohibition should be "based on the purpose for export, class of seller or purchaser, country of destination, or any other reasonable classification or basis as the President determines to be appropriate and consistent with the national interest and the purposes of this chapter."[11]

This general prohibition on crude oil exports and the exemptions to that prohibition are found in the BIS regulations on Short Supply Controls at 15 C.F.R. §754.2. The regulations provide that a license must be obtained for all exports of crude oil, including those to Canada.[12] The regulations further provide that BIS will issue licenses for certain crude oil exports that fall under one of the listed exemptions, including (i) exports from Alaska's Cook Inlet; (ii) exports to Canada for consumption or use therein; (iii) exports in connection with refining or exchange of strategic petroleum reserve oil; (iv) exports of heavy California crude oil up to an average volume not to exceed 25 million barrels per day; (v) exports that are consistent with certain international agreements; (vi) exports that are consistent with findings made by the President under certain statutes; and (vii) exports of foreign origin crude oil where, based on satisfactory written documentation, the exporter can demonstrate that the oil is not of U.S. origin and has not been commingled with oil of U.S. origin.[13] BIS administers crude oil licensing under these provisions.

Natural Gas/LNG

Section 3 of the Natural Gas Act (NGA) provides that "no person shall export any natural gas from the United States to a foreign country or import any natural gas from a foreign country without having first secured an order of the Commission authorizing it to do so."[14] This authorization is to be issued "unless, after opportunity for hearing, [the Commission] finds that the proposed exportation or importation will not be consistent with the public interest."[15] The Commission is further empowered to grant authorizations in part and to modify or place terms and conditions upon authorizations and to supplement its orders as appropriate.[16]

At the time of the NGA's enactment in 1938, the "Commission" referred to the Federal Power Commission. However, in 1977 the Federal Power Commission was dissolved and its responsibilities were transferred to the

Department of Energy (DOE) as well as the Federal Energy Regulatory Commission (FERC), an independent agency operating within DOE, pursuant to the Department of Energy Organization Act.[17] Title III of this act transferred all functions of the Federal Power Commission to DOE except for those subsequently assigned to FERC in Title IV.[18]

Title III of the DOE Organization Act thus transferred the authority to authorize natural gas imports and exports from the Federal Power Commission to DOE. Title IV provides added clarity on this point. Section 402(f) of the act specifically states that "[n]o function ... which regulates the exports or imports of natural gas or electricity shall be within the jurisdiction of [FERC] unless the Secretary assigns such functions to [FERC]."[19]

Natural gas exporting responsibilities are handled by the Office of Fossil Energy within DOE. The procedures for filing for authorization to import or export natural gas are set forth in DOE regulations found at 10 C.F.R. Part 590. The regulations establish filing requirements as well as the procedures for review of applications, including procedures that allow interested parties to participate in the process prior to the issuance of orders by DOE. The regulations also provide for an expedited filing and review process for one-time small volume imports and exports for "scientific, experimental or other non-utility gas use" without necessitating a permit.[20]

The Energy Policy Act of 1992[21] amended the NGA Section 3 generic requirement for a permit in order to export natural gas to create a more streamlined authorization process for imports from and exports to certain countries. Subsection (c) of Section 3 provides that the importation of natural gas from or exportation of natural gas to a country with which the United States has in effect "a free trade agreement requiring national treatment for trade in natural gas shall be deemed to be consistent with the public interest, and applications for such importation and exportation shall be granted without modification or delay."[22] This provision eased the authorization process for certain countries in the interest of free trade, including Canada and Mexico, the only countries with whom natural gas importation and exportation takes place via pipeline.

Section 3 of the NGA also protects the role of the states in the permitting decisions. State rights under various environmental statutes are protected with respect to both export authorization by DOE and permitting by FERC (discussed *infra*) in Section 3(d),[23] and Section 3(e) mandates the notification of relevant state authorities in order to gather their input during the process.[24]

Coal

Although the Energy Policy and Conservation Act of 1975 authorized the President to restrict coal exports,[25] the President does not appear to have exercised this authority to impose any significant export restrictions specific to coal. In fact, there have been legislative efforts aimed at expanding coal exports. For example, Section 1338 of the Energy Policy Act of 1992 directed the Secretary of Commerce to create a plan for expanding coal exports.[26] Almost all U.S. coal exports pass through ports on the East Coast or in the Gulf of Mexico,[27] so laws and regulations applicable to such facilities would potentially affect coal exports. Such laws and regulations are briefly discussed below.

EXPORT FACILITY AUTHORIZATION

The previous section of this report discusses federal authorization of the export of natural resources, not the construction and operation of export facilities. However, in many cases approval for the export facility itself also must be obtained from the federal government. This section discusses various approval requirements for different types of facilities that enable the export of oil and natural gas.

Note that, in addition to the facility approvals described below, a facility used in the export or import of fossil fuels may require additional federal approvals or authorizations. For instance, construction and operation of ports in any navigable waters in the United States are regulated by the U.S. Army Corps of Engineers (ACE). In order to construct any port facility, permits must be obtained from ACE, which will review applications to see that they are in compliance with the Clean Water Act,[28] the Rivers and Harbors Act,[29] and the Marine Protection Research and Sanctuaries Act.[30] Because coal is generally not exported via a special facility designed to transport the commodity, there are no special facility permitting requirements applicable to coal exports, but facilities through which coal (or any fossil fuel) may be exported must satisfy these generic federal requirements.

Oil Pipeline Border Crossings

Crude oil can be exported either by pipeline or via tanker or other vessel. If an oil pipeline crosses the border with Canada or Mexico, the border crossing facility must be authorized by the federal government.[31] The executive branch exercises permitting authority over the construction and operation of "pipelines, conveyor belts, and similar facilities for the exportation or importation of petroleum, petroleum products" and other products pursuant to a series of executive orders. This authority has been vested in the U.S. State Department since the promulgation of Executive Order 11423 in 1968.[32] Executive Order 13337 amended this authority and the procedures associated with the review, but did not substantially alter the exercise of authority or the delegation to the Secretary of State in Executive Order 11423.[33]

Executive Order 11423 provides that, except with respect to cross-border permits for electric energy facilities, natural gas facilities, and submarine facilities:

> The Secretary of State is hereby designated and empowered to receive all applications for permits for the construction, connection, operation, or maintenance, at the borders of the United States, of: (i) pipelines, conveyor belts, and similar facilities for the exportation or importation of petroleum, petroleum products, coal, minerals, or other products to or from a foreign country; (ii) facilities for the exportation or importation of water or sewage to or from a foreign country; (iii) monorails, aerial cable cars, aerial tramways and similar facilities for the transportation of persons or things, or both, to or from a foreign country; and (iv) bridges, to the extent that congressional authorization is not required.[34]

Executive Order 13337 designates and empowers the Secretary of State to "receive all applications for Presidential Permits, as referred to in Executive Order 11423, as amended, for the construction, connection, operation, or maintenance, at the borders of the United States, of facilities for the exportation or importation of petroleum, petroleum products, coal, or other fuels to or from a foreign country."[35] Executive Order 13337 further provides that after consideration of the application and comments received:

> If the Secretary of State finds that issuance of a permit to the applicant would serve the national interest, the Secretary shall prepare a permit, in such form and with such terms and conditions as the national

> interest may in the Secretary's judgment require, and shall notify the officials required to be consulted ... that a permit be issued.[36]

Thus, the Secretary of State is directed by the order to authorize those border crossing facilities that the Secretary has determined would "serve the national interest."

Note that the source of the executive branch's permitting authority is not explicitly stated within the executive orders. Powers exercised by the executive branch are authorized by legislation or are inherent presidential powers based in the Constitution. Executive Order 11423 does not reference any statute or constitutional provision as the source of its authority, although it does state that "the proper conduct of foreign relations of the United States requires that executive permission be obtained for the construction and maintenance" of border crossing facilities.[37] Executive Order 13337 refers only to the "Constitution and the Laws of the United States of America, including Section 301 of title 3, United States Code."[38] Section 301 of Title 3 provides that the President is empowered to delegate authority to the head of any department or agency of the executive branch. Courts that have addressed the legitimacy of this exercise of authority have found that it is a legitimate exercise of "the President's constitutional authority over foreign affairs and his authority as Commander in Chief."[39]

Natural Gas Pipeline Border Crossings

As discussed above, Executive Orders 11423 and 13337 explicitly exclude cross-border natural gas pipelines (among others) from their reach. Instead, permitting for these facilities is addressed in Executive Order 10485, which governs the issuance of Presidential Permits for natural gas facilities.[40] Executive Order 10485 designates and empowers the now-defunct Federal Power Commission:

1) To receive all applications for permits for the construction, operation, maintenance, or connection, at the borders of the United States, of facilities for the transmission of electric energy between the United States and a foreign country.
2) To receive all applications for permits for the construction, operation, maintenance, or connection, at the borders of the United States, of

facilities for the exportation or importation of natural gas to or from a foreign country.

3) Upon finding the issuance of the permit to be consistent with the public interest, and, after obtaining the favorable recommendations of the Secretary of State and the Secretary of Defense thereon, to issue to the applicant, as appropriate, a permit for such construction, operation, maintenance, or connection. The Secretary of Energy shall have the power to attach to the issuance of the permit and to the exercise of the rights granted thereunder such conditions as the public interest may in its judgment require.[41]

In many ways, this authority resembles the authority over oil pipelines granted to the State Department in Executive Orders 11423 and 13337. However, as mentioned above, Executive Orders 11423 and 13337 do not describe the source of the executive branch permitting authority granted by the orders. Judicial opinions strongly suggest the permitting authority is an exercise of the President's "inherent constitutional authority to conduct foreign affairs."[42] By contrast, Executive Order 10485 cites federal statutes which may at least partially form the basis for the permitting authority granted to the DOE by the order. The order states that "section 202(e) of the Federal Power Act, as amended .. requires any person desiring to transmit any electric energy from the United States to a foreign country to obtain an order from the Federal Power Commission authorizing it to do so" and that "section 3 of the Natural Gas Act ... requires any person desiring to export any natural gas from the United States to a foreign country or to import any natural gas from a foreign country to the United States to obtain an order from the Federal Power Commission authorizing it to do so." These appeals to statutory authority should be considered and possibly addressed in any legislation seeking to amend the current Presidential Permit process for border crossings for energy facilities.

The Department of Energy Organization Act of 1977[43] eliminated the Federal Power Commission and transferred its functions to either the newly created DOE or the FERC, an independent regulatory agency within DOE. Section 402(f) of that act specifically reserved import/export permitting functions for DOE rather than FERC. As a result, DOE took over the FPC's Presidential Permit authority for border crossing facilities under Executive Order 10485 pursuant to the act. The authority to issue Presidential Permits for natural gas pipeline border crossings was subsequently transferred to FERC in 2006 via DOE Delegation Order No. 00-004.00A.[44]

LNG Export Terminals

Section 3(e) of the NGA, adopted in Section 311 of the Energy Policy Act of 2005,[45] assigns the "exclusive authority to approve or deny an application for the siting, expansion or operation of a Liquefied Natural Gas (LNG) terminal" to FERC.[46] Section 3 designates FERC as the "lead agency for the purposes of coordinating all applicable Federal authorizations" and for complying with federal environmental requirements.[47] Section 3(e) also directs FERC to promulgate regulations for pre-filing of LNG import terminal siting applications and directs FERC to consult with designated state agencies regarding safety in considering such applications.[48]

FERC implements its authority over onshore LNG terminals through the agency's regulations at 18 C.F.R. §153. These regulations detail the application process and requirements under Section 3 of the NGA. The process begins with a pre-filing, which must be submitted to FERC at least six months prior to the filing of a formal application. The pre-filing procedures and review processes are set forth at 18 C.F.R. §157.21. Once the pre-filing stage is completed, a formal application may be filed. FERC's formal application requirements include detailed site engineering and design information, evidence that a facility will safely receive or deliver LNG, and delineation of a facility's proposed location.[49] The regulations also require LNG facility builders to notify landowners who would be affected by the proposed facility.[50] To facilitate natural gas infrastructure projects, which includes LNG projects, FERC has adopted rules to provide "blanket certificates" that provide authorization to interstate pipelines to improve or upgrade existing facilities or construct certain new facilities pursuant to a streamlined process.[51]

World Trade Organization—General Agreement on Tariffs and Trade

The Marrakesh Agreement Establishing the World Trade Organization (WTO) contains the agreements relating to international trade that are binding for all WTO Members. Although there is no specific agreement relating to trade in energy products, such as liquefied natural gas, coal, or oil, the trade in these products is regulated under the General Agreement on Tariffs and Trade (GATT). Several of these sections could potentially impact a nation's ability to limit or restrict fossil fuels.

Article I—Most Favored Nation Treatment

Article I of the GATT 1994 requires that "any advantage, favour, privilege or immunity granted by any [WTO Member] to any product originating in or destined for any other country shall be accorded immediately and unconditionally to the like product originating in or destined for the territories of all other [WTO Members]."[52] Article I applies to all rules and formalities in connection with importation and exportation.[53] This broad category of rules and formalities appears likely to include prerequisites for exportation such as licensing requirements or other preliminary measures.[54] More favorable treatment given to *imports* from particular countries in the context of *import* licensing requirements has been held to confer an advantage within the meaning of Article I.[55]

Generally, this means that as soon as the United States provides for certain treatment of fossil fuel exports to one country, the United States has to treat exports to all other WTO Members in the same fashion. A licensing regime that provided for more favorable treatment for exports of fossil fuels to some countries, but subjected other WTO countries to a slower process could potentially be inconsistent with Article I of the GATT.

However, there are exceptions to the Most Favored Nation Treatment requirements for Free Trade Agreements (FTA). Article XXIV of GATT 1994 allows countries to provide more favorable treatment to countries with which they have established an FTA.[56] In order to qualify for the Article XXIV exception, the FTA must meet certain requirements outlined in the Article. Most notably, the free trade agreement must eliminate duties—such as tariffs—and restrictions on commerce between the parties to the agreement for "substantially all the trade in products originating in those territories."[57] Therefore, in order for an agreement to qualify, it is likely that the FTA would have to cover more than just energy products flowing between the two territories. However, if the countries have a qualifying FTA, more favorable treatment towards energy products between those countries could be included in that FTA without violating the GATT.

Article XI—Export Restrictions

Article XI of the GATT covers import and export restrictions. Article XI:1 of the GATT bars the institution or maintenance of quantitative restrictions on exports to any WTO Member's territory.[58] Quantitative restrictions limit the amount of a product that may be exported—common examples are embargoes, quotas, minimum export prices, and certain export licensing requirements. Under Article XI, duties, taxes, and other charges are the only GATT-consistent methods of restricting exports.[59] Any government action that expressly precludes the exportation of certain goods is inconsistent with the GATT.

Although there are few WTO panel decisions on export bans, panels have consistently found that *import* bans implemented through licensing systems violate Article XI.[60] This jurisprudence can be expected to inform any WTO panel decision on the GATT-consistency of export bans and licensing.[61] WTO Panel decisions have also held that "discretionary" or "non-automatic" licensing requirements are prohibited under Article XI—therefore, a licensing program that gives discretion to an agency to deny an export license to potential exporters on the basis of vague or unspecified criteria would violate Article XI.[62] Moreover, a GATT panel held that export licensing practices that cause delays in issuing licenses may be a restraint of exports that is inconsistent with Article XI.[63]

Articles VI, XVI, and the Agreement on Subsidies and Countervailing Measures—Export Restraints as Actionable Subsidies

A fossil fuel export licensing regime that restricts exports could have the effect of keeping domestic prices of fossil fuels lower than they otherwise would be. This raises the question of whether such a licensing program could be considered an actionable subsidy to downstream users of the fossil fuels such as members of the petrochemical industry. Under the GATT 1994 and the Agreement on Subsidies and Countervailing Measures (SCM Agreement), an actionable subsidy may be the subject of countervailing measures or challenge before a panel by a WTO Member when the subsidy adversely affects the interests of that member.[64] Adverse effects might result if export restraints on fossil fuels lead to lower input costs for downstream manufacturers that use the fuels, giving the manufacturers' products a competitive edge over the products of the other members' manufacturers in domestic or foreign markets.

The SCM Agreement defines a "subsidy" as "a financial contribution by a government or any public body within the territory of a Member" that confers a benefit.[65] Under the agreement, one way that a "financial contribution" may occur is when a government directs a private body to sell goods to a domestic purchaser.[66] In *U.S.—Measures Treating Export Restraints as Subsidies*, the United States Trade Representative (USTR) argued before a WTO panel that a government's restriction on exports could be considered "functionally equivalent" to that government directing private parties to sell a good to domestic purchasers.[67] The USTR argued that this resulted in a subsidy to downstream producers that used the good as an input in their production processes.[68] The panel rejected this argument, stating that although a restriction on exports of a good may result in lower prices for domestic users of that good, the restriction was not an explicit command or direction by the government to private parties to sell the good within the meaning of the SCM Agreement.[69] This ruling suggests that future panels may be reluctant to find that a restriction on exports or a similar government intervention in a market is a "financial contribution" by a government. Thus, it seems unlikely that licensing procedures could constitute a subsidy under WTO rules, even if they lead to restrictions on exports.

Articles XX and XIII—General Exceptions

Article XX of the GATT provides for certain exceptions that a member country may invoke if it is found to be in violation of any GATT obligations. In order for the defense to be successful, the member country must show that its action fits under one of these general exceptions and that it satisfies Article XX's opening clauses, known as the "chapeau."[70] When dealing with trade in energy products, a country will most likely use the exceptions under Article XX(b) or XX(g). A country may justify a GATT inconsistent practice under Article XX(b) if the practice in question is "necessary to protect human, animal, or plant life or health."[71] Article XX(g) may permit otherwise GATT inconsistent measures that "relat[e] to the conservation of exhaustible natural resources if such measures are made effective in conjunction with restrictions on domestic production or consumption."[72] If a WTO Member invokes an Article XX exception to the application of any quantitative export restrictions, Article XIII requires that those export restrictions must be administered in a non-discriminatory manner—that is the restrictions must comport with the most-favored nation treatment discussed above.[73]

Article XXI—Security Exceptions

Restrictions on fossil fuels for reasons of international or domestic security that would otherwise violate the GATT 1994 may potentially be justified under the broadly worded exception for essential security interests contained in Article XXI.[74] One paragraph of this article allows a member to take "any action which it considers necessary for the protection of its essential security interests ... taken in time of war or other emergency in international relations."[75]

Because there is a lack of WTO case law on Article XXI, it is necessary to consult the history of the exception's use under the General Agreement on Tariffs and Trade 1947 (GATT 1947). The GATT 1947 has been incorporated into the GATT 1994. Under the GATT 1947, the contracting parties broadly interpreted the concept of an "emergency."[76] Thus, each party was basically the judge of what constituted an emergency in international relations. Measures that parties have sought to justify under Article XXI have included trade embargoes, import quotas, and suspensions of tariff concessions.[77] Parties have pointed to both potential and actual dangers as "emergencies" supposedly justifying these typically GATT inconsistent measures.[78]

With respect to who determines whether a WTO Member's use of the exception is valid, the United States has previously taken the position that Article XXI is "self-judging."[79] That is, each member invoking Article XXI is the judge of whether its use of the exception is valid. As a result, there is currently no WTO case law on the use of Article XXI. However, some scholars have speculated that, in the future, a WTO panel or the Appellate Body may decline to defer to a WTO Member's judgment about when its use of Article XXI is appropriate.[80] One of these international quasi-judicial bodies may instead decide to more carefully scrutinize a member's use of the exception. For example, a panel may consider whether there is an emergency in international relations justifying national security screening for exports of fossil fuels to certain countries but not others.

NAFTA and Other Free Trade Agreements

In addition to the GATT, the United States is party to numerous FTAs. It is beyond the scope of this report to discuss fully the provisions of each FTA signed by the United States. However, as an example, several FTAs require national treatment for trade in natural gas. These include FTAs with Australia,

Bahrain, Canada, Chile, Colombia, Dominican Republic, El Salvador, Guatemala, Honduras, Jordan, Mexico, Morocco, Nicaragua, Oman, Panama, Peru, Republic of Korea and Singapore.[81] The FTAs with Costa Rica and Israel do not require national treatment for trade in natural gas.[82]

As a further example, under the North American Free Trade Agreement (NAFTA) the United States has certain obligations related to energy trade with Mexico and Canada. Chapter 6 of NAFTA deals with "Energy and Basic Petrochemicals." Chapter 6 reconfirms the Parties' obligations under the GATT and imposes additional obligations on the Parties, such as certain requirements for export taxes.[83] NAFTA also imposes barriers to invoking some of the general exceptions to the GATT.[84] For example, a country may only invoke the "conservation of exhaustible natural resources" exception if it does not result in a higher price for exports than for domestic consumption of the energy products.[85] These additional obligations illustrate that compliance with FTAs must be considered when establishing export regulations for energy products.

PENDING LEGISLATION

S. 192, the Expedited LNG for American Allies Act of 2013, was introduced in the Senate on January 31, 2013, by Senator John Barrasso. An identical bill, H.R. 580, was introduced in the House of Representatives by Representative Michael Turner. The bills would amend section 3 of the NGA to provide that expedited approval of LNG exports would be granted to four different categories of foreign countries: (1) nations for which there is in effect a free trade agreement (FTA) requiring national treatment for trade in natural gas; (2) a member country of the North Atlantic Treaty Organization (NATO); (3) Japan, so long as the Treaty of Mutual Cooperation and Security of January 19, 1960, between Japan and the United States remains in effect; and (4) "any other foreign country if the Secretary of State, in consultation with the Secretary of Defense, determines that exportation of natural gas to that foreign country would promote the national security interests of the United States."[86]

At least two other bills that pertain to the export of natural gas have been introduced. H.R. 1189, the American Natural Gas Security and Consumer Protection Act, was introduced in the House by then-Representative Ed Markey. The bill would amend the NGA to require the Secretary of Energy to develop regulations for determining whether an export of natural gas from the United States to a foreign country is in the public interest for the purposes of issuing an export authorization.[87] Under the regulations, the public interest determination would have to be made after the secretary's consideration of several factors, including the energy security of the United States; the ability of the United States to reduce greenhouse gas emissions; and an environmental impact statement issued under the National Environmental Policy Act that analyzes the impact of extraction of exported natural gas on the environment in communities where the gas is extracted.[88] H.R. 1191, the Keep American Natural Gas Here Act, was also introduced in the House by then-Representative Ed Markey. Among other things, it would provide that the Secretary of the Interior could accept bids on new oil and gas leases of federal lands (including submerged lands) only from bidders certifying that all natural gas produced pursuant to such leases would be sold only in the United States.[89]

With regard to oil, H.R. 1190, the Keep America's Oil Here Act, was introduced in the House by then-Representative Ed Markey. The bill would provide that the Secretary of the Interior could accept bids on new oil and gas leases of federal lands (including submerged lands) only from bidders certifying that oil produced pursuant to such leases, and any refined petroleum products produced from that oil, would be sold only in the United States.[90] The bill would allow the President to waive this requirement for a lease in certain circumstances, including when a waiver is necessary under an international agreement.[91] In addition, S. 435, the American Oil for American Families Act of 2013, was introduced in the Senate by Senator Robert Menendez. The bill would ban the export of crude oil or refined petroleum products derived from federal lands (including land on the Outer Continental Shelf).[92]

CONCLUSION

Recent advances in natural gas exploration and production technology have led to a newfound interest in the possibility of expanding US fossil fuel exports. Such exports, and the facilities needed to conduct export operations, are subject to a panoply of federal laws and regulations. These include the authorizations required by the Natural Gas Act, a generic ban on crude oil exports, and various laws and regulations applicable to construction and operation of export facilities. Currently, any party wishing to export fossil fuels must comply with these laws and regulations.

Under international trade rules, restrictions on exports of fossil fuels could potentially be difficult to reconcile with Articles I and XI of the GATT 1994. Article XXI, the exception for essential security interests, may be cited in order to justify potential violations of GATT Articles I and XI. The United States has traditionally considered this exception to be self-judging. However, it is possible that a panel or the Appellate Body might scrutinize the United States' use of the exception.

Article XX of the GATT provides additional exceptions that a member country may invoke if it is found to be in violation of any GATT obligations. However, Article XIII requires that if an otherwise GATT inconsistent measure is permitted to remain in force due to an Article XX exception, the measure must be administered in a non-discriminatory manner. Export restrictions that treat WTO Members differently would appear not to satisfy the non-discriminatory requirements of Article XIII.

End Notes

[1] For more information about the potential for natural gas exports, see CRS Report R42074, *U.S. Natural Gas Exports: New Opportunities, Uncertain Outcomes*, by Michael Ratner et al.

[2] For more information about the potential for crude exports, see CRS Report R42465, *U.S. Oil Imports and Exports*, by Robert Pirog.

[3] Export transactions that fall within the exclusive jurisdiction of another federal agency are not subject to the EAR. 15 C.F.R. Part 734. In some cases, more than one federal agency may be responsible for exercising oversight over the export of a particular item.

[4] 15 C.F.R. Part 732.

[5] *Id.*

[6] *Id.*

[7] 15 C.F.R. Parts 30 and 758. The Bureau of Census uses the SED or AES to compile trade statistics. 15 C.F.R. §758.1. BIS uses the records for export control purposes. *Id.* Circumstances in which an SED or AES record is required to be submitted for the export of items subject to the EAR include when the items are destined for certain countries; when the export of the items requires submission of a license application under the EAR; and when the value of the exported commodities classified under a single Schedule B Number (or Harmonized Tariff Schedule number) exceeds $2,500. *Id.* Certain exceptions may apply. *See id.*

[8] U.S. Department of Commerce, A Basic Guide to Exporting 75 (1998).

[9] P.L. 94-163.

[10] 42 U.S.C. §6212(b)(1).

[11] *Id.* at §6212(b)(2).

[12] 15 C.F.R. §754.2(a).

[13] *Id.* at §754.2(b).

[14] 15 U.S.C. §717b(a). This prohibition on natural gas exports without a determination from the executive branch that the national or public interest echoes the language found in the Energy Policy and Conservation Act of 1975, discussed above. The DOE regulations at 10 CFR. Part 590 cite to both acts as authority for the regulations governing export permitting.

[15] *Id.*

[16] *Id.*
[17] P.L. 95-91.
[18] *Id.* at §301.
[19] 42 U.S.C. §7172(f).
[20] 10 C.F.R. §590.208.
[21] P.L. 102-486.
[22] 15 U.S.C. §717b(c).
[23] *Id.* at §717b(d).
[24] *Id.* at §717b(e)(2).
[25] 42 U.S.C. §6212(a).
[26] 42 U.S.C. §13367.
[27] Energy Information Admin, Quarterly Coal Exports October-December 2012 (March 2013), Table 13: U.S. Coal Exports by Customs District, available at http://www.eia.gov/coal /production/quarterly/pdf/t13p01p1.pdf.
[28] 33 U.S.C. §1344.
[29] 33 U.S.C. §403.
[30] 33 U.S.C. §§1401 et. seq.
[31] For tankers or other vessels conveying oil, the use of such facilities for exportation, in and of itself, does not require a permit akin to that required for oil pipelines that cross international borders. However, oil tankers or other such vessels must comply with other generally applicable export requirements, which are discussed elsewhere in this report.
[32] Exec. Order No. 11423, *Providing for the performance of certain functions heretofore performed by the President with respect to certain facilities constructed and maintained on the borders of the United States*, 33 Fed. Reg 11741. (August 20, 1968).
[33] Exec. Order No. 13337, *Issuance of Permits With Respect to Certain Energy-Related Facilities and Land Transportation Crossings on the International Boundaries of the United States*, 69 Fed. Reg. 25299 (May 5, 2004).
[34] Exec. Order No. 11423, 33 Fed. Reg. at 11741.
[35] Exec. Order No. 13337, 69 Fed. Reg. at 25299.
[36] *Id.* at 25230.
[37] 33 Fed. Reg. at 11741.
[38] 69 Fed. Reg. at 25299.
[39] Sierra Club v. Clinton, 689 F. Supp. 2d 1147, 1162 (D. Minn. 2010).
[40] Exec. Order No. 10485, *Providing for the performance of certain functions heretofore performed by the President with respect to electric power and natural gas facilities located on the borders of the United States*, 18 Fed. Reg. 5397 (Sept. 3, 1953).
[41] *Id.*
[42] Sisseton-Wahpeton Oyate v. U.S. Department of State, 659 F. Supp. 2d 1071, 1081 (D.S.D. 2009).
[43] P.L. 95-91, 42 U.S.C. §4101 note.
[44] Available at http://www.ferc.gov/industries/electric/indus-act/siting/doe-delegation.pdf.
[45] P.L. 109-58.
[46] 15 U.S.C. §717b(e). Gas must be converted to LNG for export by means other than pipeline.
[47] *Id.*
[48] *Id.*
[49] 18 C.F.R. §153.8.
[50] 18 C.F.R. §157.6d.
[51] 18 C.F.R. §§157.201-157.218.
[52] General Agreement on Tariffs and Trade 1994, Art. I:1 (hereinafter GATT 1994).

[53] *Id.*

[54] *See* Panel Report, *U.S.—Certain Measures Affecting Imports of Poultry from China*, paras. 7.407, 7.410, WT/DS392/R (September 29, 2010) ("We conclude that 'in connection with importation' as used in Article I, not only encompasses measures which directly relate to the process of importation but could also include those measures ... which relate to other aspects of the importation of a product or have an impact on actual importation."). The same reasoning could apply to measures that have an impact on actual exportation, such as licensing requirements.

[55] Panel Report, *EC—Regime for the Importation, Sale, and Distribution of Bananas,* 7.193, WT/DS27/R/USA (May 22, 1997).

[56] GATT 1994, Art. XXIV.

[57] *Id.*

[58] GATT 1994, Art. XI:1 ("No prohibitions or restrictions other than duties, taxes or other charges, whether made effective through quotas, import or export licences [sic] or other measures, shall be instituted or maintained by any contracting party on the ... exportation or sale for export of any product destined for the territory of any other contracting party.").

[59] *Id.*

[60] *See* Panel Report, *Brazil—Measures Affecting Imports of Retreaded Tyres*, WT/DS332/R (June 12, 2007) (holding a licensing system to be in violation of Article XI when a person would be ineligible to import tires based on where those tires came from); Panel Report, *India Quantitative Restrictions on Imports of Agricultural, Textile and Industrial Products*, WT/DS90/R (April 6, 1999).

[61] *See* Wen-Chen Shih, *Energy Security, GATT/WTO, and Regional Agreements*, 49 Nat. Res. J. 433, 451 (2009) (noting that it is likely that "the jurisprudence concerning quantitative restrictions on import in the interpretation and application of Article XI:1 also applies to exports.").

[62] *See* Panel Report, *China—Measures Related to the Exportation of Various Raw Materials*, WT/DS394/R (July 5, 2011) (holding that vague export licensing criteria allowed for too much discretion in granting licenses and that they were therefore in violation of Article XI).

[63] Panel Report, *Japan—Trade in Semi-Conductors* (May 4, 1988) GATT B.I.S.D. (35th Supp.), 31.

[64] GATT 1994, Arts. VI, XVI; SCM Agreement, Arts. V, VII, XI, XIX.

[65] SCM Agreement, Art. I.

[66] SCM Agreement, Art. I(a)(1)(iv).

[67] Panel Report, *United States—Measures Treating Export Restraints as Subsidies*, 8.22, WT/DS194/R (June 29, 2001).

[68] *Id.* at paras. 5.36, 5.48-.51.

[69] *Id.* at 8.42-.44.

[70] The "chapeau" requires, for example, that measures falling under these exceptions shall not be a disguised restriction on international trade. GATT 1994, Art. XX.

[71] GATT 1994, Art. XX(b). It is worth noting that the "necessary" requirement is a rather high standard to meet. Appellate Body Report, *Brazil—Measures Affecting Imports of Retreaded Tyres*, 150, WT/DS332/AB/R (December 3, 2007).

[72] GATT 1994, Art. XX(g). Although "relating to" may be an easier standard to meet when relying on the exception, the measure in question must also operate in conjunction with domestic restrictions.

[73] GATT 1994, Art. XIII.

[74] GATT 1994, Art. XXI.

[75] GATT 1994, Art. XXI(b)(iii).

[76] *See* GATT Analytical Index—Guide to GATT Law and Practice 602-05 (6th ed. 1995).

[77] *Id.* at 602-05.

[78] *Id.* at 600.

[79] Dapo Akande & Sope Williams, *International Adjudication on National Security Issues: What Role for the WTO?*, 43 Va. J. Int'l L. 365, 375-76 (2003).

[80] *Id.* at 383-84.

[81] Department of Energy, How to Obtain Authorization to Import and/or Export Natural Gas and LNG, http://energy.gov/fe/how-obtain-authorization-import-andor-export-natural-gas-and-lng.

[82] *Id.*

[83] North American Free Trade Agreement, Arts. 603, 604 [hereinafter NAFTA].

[84] NAFTA, Art. 605.

[85] *Id.*

[86] S. 192, §2.

[87] H.R. 1189, §2.

[88] *Id.*

[89] H.R. 1191, §2.

[90] H.R. 1190, §3.

[91] *Id. §4.*

[92] S. 435, §2.

In: United States Fossil Fuel Exports
Editor: Mitchell P. Henning

ISBN: 978-1-63117-198-7

Chapter 2

U.S. Natural Gas Exports: New Opportunities, Uncertain Outcomes*

Michael Ratner, Paul W. Parfomak, Ian F. Fergusson and Linda Luther

Summary

As estimates for the amount of U.S. natural gas resources have grown, so have the prospects of rising U.S. natural gas exports. The United States is expected to go from a net importer of natural gas to a net exporter by 2020. Projects to export liquefied natural gas (LNG) by tanker ship have been proposed—cumulatively accounting for about 41.4% of current gross U.S. natural gas production—and are at varying stages of regulatory approval. Projects require federal approval under Section 3 of the Natural Gas Act (15 U.S.C. §717b), with the U.S. Department of Energy's Office of Fossil Energy and the Federal Energy Regulatory Commission being the lead authorizing agencies. Pipeline exports, which accounted for 98% of all exports of U.S. produced natural gas in 2012, are also likely to continue rising.

What effect exporting natural gas will have on U.S. prices is a central question in the debate over whether to export. A significant rise in U.S. natural gas exports would likely put upwards pressure on domestic prices,

* This is an edited, reformatted and augmented version of a Congressional Research Service publication, No. R42074, dated September 17, 2013.

but the magnitude of any rise is currently unclear. There are numerous factors that will affect prices: export volumes, economic growth, differences in local markets, and government regulations, among others. With recent natural gas prices relatively low compared to global prices and historically low for the United States, producers are looking for new markets for their natural gas. Producers contend that increased exports will not raise prices significantly as there is ample supply to meet domestic demand, and there will be the added benefits of increased revenues, trade, and jobs, and less flaring. Consumers of natural gas, who are being helped by the low prices, fear prices will rise if natural gas is exported.

Electric power generation represents potentially the greatest increase in natural gas consumption in the U.S. economy, primarily for environmental reasons. Natural gas emits much less carbon dioxide and other pollutants than coal when combusted. Other types of consumption are not likely to increase natural gas demand domestically for a long time. Use in the transportation sector to displace oil is likely to be small because expensive new infrastructure and technologies would be required. There is discussion of a possible revival of the U.S. petrochemicals sector, but the potential extent of a change is unclear.

Getting natural gas to markets where it can be consumed, whether domestically or internationally, may be the industry's biggest challenge. Infrastructure constraints, environmental regulations, and other factors will influence how the market adjusts to balance supply and demand.

Environmental groups are split regarding natural gas use, with some favoring increased use to curb emissions of certain pollutants, while others oppose expanded use of natural gas because it is not as clean as renewable forms of energy, such as wind or solar. The use of hydraulic fracturing to produce shale gas has also raised concerns among environmental groups particularly concerned with its possible impacts on water quality.

The possibility of a significant increase in U.S. natural gas exports will factor into ongoing debates on the economy, energy independence, climate change, and energy security. As the proposed projects continue to develop, policymakers are likely to receive more inquiries about these projects. Proposals to expedite and expand LNG exports have been raised in the 113th Congress, including in S. 192 and H.R. 580. Two other bills, H.R. 1189 and H.R. 1191, would reform the DOE's process for determining the public interest regarding LNG exports and prohibit exports of natural gas produced on federal lands.

INTRODUCTION: THINGS HAVE CHANGED

The United States has exported natural gas for close to 100 years, but has generally exported less than it has imported (mostly from Canada).[1] Within the next five years, the United States may become a much larger exporter of natural gas, particularly liquefied natural gas (LNG), for the first time. Increased development of U.S. natural gas resources—primarily shale gas—along with low domestic prices in recent years and idle LNG import infrastructure, have driven the change in the U.S. position. As recently as the mid-to-latter 2000s, the United States was projected to be a growing natural gas importer. However, imports have been declining since 2005, while exports have been climbing.

U.S. natural gas exports require federal approval pursuant to Section 3 of the Natural Gas Act (NGA) (15 U.S.C. §717b), with the U.S. Department of Energy's (DOE) Office of Fossil Energy and the Federal Energy Regulatory Commission (FERC) being the lead authorizing agencies. Historically, exports have been primarily via pipeline to Mexico and eastern Canada, but natural gas companies are now considering exporting greater quantities of U.S. LNG by tanker ship to a number of other countries. From 1969 to 2012, the United States exported Alaskan LNG almost exclusively to Japan, but the volumes of those shipments have been relatively small. Furthermore, Alaska's natural gas market has been, and continues to be, isolated from the rest of the United States. The prospect of the United States supplying a global market with large quantities of LNG from the lower 48 states raises concerns in Congress, particularly about a potential rise in what U.S. consumers pay for natural gas and effects on the economy.[2]

Developers of natural gas export projects and natural gas producers argue that domestic gas prices will not rise much, or significantly, if U.S. natural gas exports increase because the United States has ample gas resources to meet domestic demand. Further, they argue there will be economic benefits such as increased employment and an improved trade balance. Other stakeholders disagree, fearing that such exports could cause domestic natural gas prices to rise thereby hurting the economy. Some environmental groups fear that increased exports will cause more shale gas production, which they are against. DOE has used the results of two studies it commissioned about the impact on domestic gas prices of exports and the effect on the U.S. economy to justify the approval of projects.

U.S. natural gas prices are lower than those in other international markets, partly because of the competitive nature of the U.S. market. Nevertheless, natural gas prices within the United States vary by regional market because of transportation limitations, access to supplies, and differences in demand. As new volumes of shale gas are developed, these supplies will seek markets where little or no natural gas production has existed in the past. Over time, the U.S. natural gas market will reconfigure itself to balance supply and demand regionally and nationally. But getting new natural gas supplies to market may be an ongoing challenge for the industry, whether within the United States or abroad. Infrastructure constraints, such as the availability of pipelines, environmental regulations, and other regulatory requirements will play a part in how the natural gas market adjusts. Exports of natural gas either by pipeline or as LNG will be a factor as the market seeks balance, especially on a regional basis. Hence the potential export of more U.S. natural gas may have economic effects that vary significantly from region to region, and regional impacts may diverge from impacts on the nation as a whole.

Other issues have also been raised regarding natural gas exports. Environmental groups are divided on the desirability of greater use of natural gas at home and abroad. Advocates see it as reducing emissions compared to other hydrocarbons, whereas opponents point out that natural gas still emits carbon dioxide and other pollutants. Concerns about contamination of water supplies during gas production have been raised because of the use of hydraulic fracturing ("fracking"), the technique for extracting shale gas which uses water, sand, and chemicals to create fissures in shale, allowing the trapped natural gas to be cost-effectively extracted.[3] Other groups want to see greater use of natural gas in the U.S. economy for economic and national security concerns before it is exported overseas.[4]

The prospect of growing U.S. natural gas exports, particularly LNG, portends to be a factor as Congress debates the economy, energy independence, climate change, and energy security. Bills to expedite and expand LNG exports have been introduced in the 113th Congress, including S. 192 and H.R. 580. Two other bills, H.R. 1189 and H.R. 1191, would reform the DOE's process for determining the public interest regarding LNG exports and prohibit exports of natural gas produced on federal lands. This report examines what has changed in the U.S. natural gas market and the prospects and implications of the United States becoming a significant net natural gas exporter.

BACKGROUND: NATURAL GAS EXPORTS ARE NOT NEW

Heading into the 2000s, the United States was expected to be a growing importer of natural gas because domestic production was declining and demand was rising (see Figure 1). The U.S. Energy Information Administration (EIA) in its 1999 Annual Energy Outlook forecasted that net natural gas imports would grow between 1997 and 2020 from 12.9% of consumption to 15.5%, based on consumption growing faster than production.[5] To accommodate the potential increase in imports, five new LNG import terminals were built by industry in the latter half of the 2000s and some existing facilities were re-commissioned and expanded. The United States currently has LNG import capacity of almost 14 billion cubic feet per day (bcf/d) or over five trillion cubic feet (tcf) per year. However, higher domestic production—mainly from shale gas development—has made imports largely unnecessary, leaving existing import capacity mostly idle. (See Table B-1 for the U.S. supply and demand balance.) In its Annual Energy Outlook 2013 Early Release, EIA projects that the United States will be a net LNG exporter by 2016 and a net natural gas exporter by 2020.[6]

The abundance of new domestic natural gas supplies shifted industry interest from building LNG import terminals to constructing LNG export terminals. As of September 2013, there have been 31 applications for permits to construct liquefaction facilities at existing LNG import terminals (also known as regasification facilities) or for new LNG export facilities in order to export domestically produced natural gas as LNG. (See Table 1.) The total capacity proposed is approximately 33.8 bcf/d. Additionally, seven companies have active permits and one other has been approved to re-export LNG cargos (take in foreign cargos, hold in storage, and then reload onto LNG tankers to go to foreign markets) from import terminals. Increased pipeline exports to Canada[7] and Mexico may also continue to rise if their domestic demand continues to increase.

Low U.S. natural gas prices relative to other international markets have spurred interest in exporting U.S. produced natural gas (see Figure 2). What effect exporting natural gas will have on U.S. prices and the overall U.S. economy are central questions in the debate over whether to export. DOE commissioned two studies as part of its evaluation of whether LNG exports to nonfree-trade-agreement (non-FTA) countries are in the public interest (see "DOE and the Public Interest Determination").

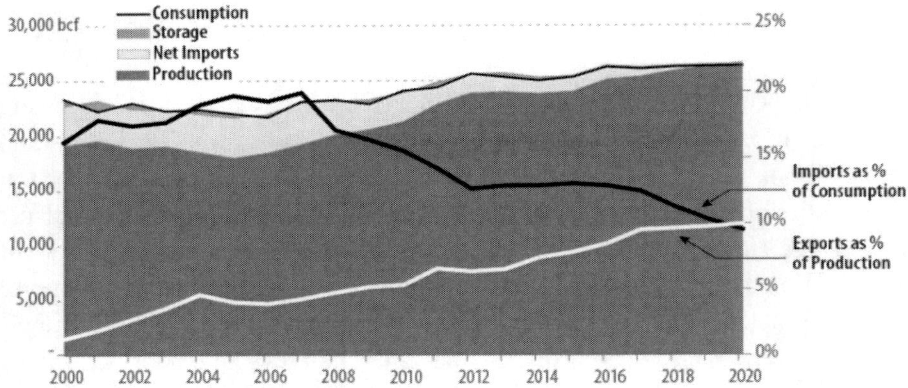

Source: EIA natural gas databases, http://www.eia.doe.gov/naturalgas/data.cfm, and EIA's Annual Energy Outlook 2013 Early Release, Natural Gas Section, reference case, http://www.eia.gov/oiaf/aeo/tablebrowser/.

Notes: Consumption equals Production, Net Imports, and Storage, but because of negative storage numbers for injection of natural gas into facilities the Consumption line in the graph does not exactly align with the sum of its parts. Exports of natural gas include the re-export of LNG cargos that comprise third-country natural gas supplies. Units are billion cubic feet (bcf) per year. Historical data are from 2011 and earlier.

Figure 1. U.S. Natural Gas Production, Consumption, and Trade; Historical and Projected Data.

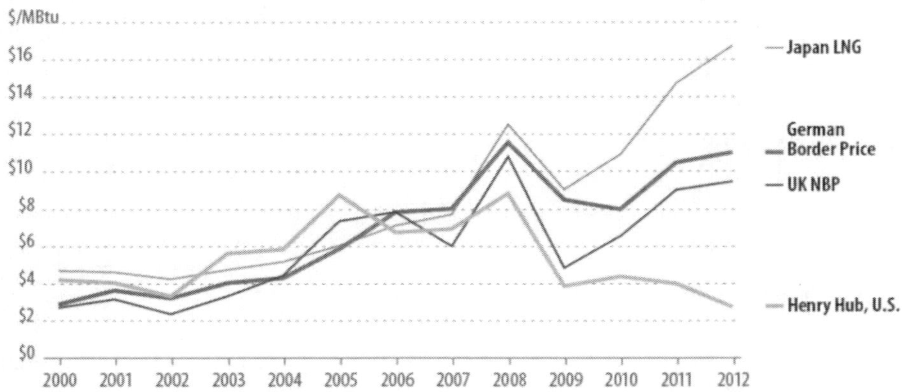

Source: BP Statistical Review of World Energy, 2013, June 2013, p. 27. http://www.bp.com/sectionbodycopy.do? categoryId=7500&contentId=7068481.

Notes: The German Border Price is a proxy for European oil indexed prices. Units = U.S. dollar per million British thermal units (MBtu). NBP is the market price in the United Kingdom referred to as the National Balancing Point.

Figure 2. Select Global Natural Gas Prices; Nominal dollars.

The most recent EIA projections show the United States becoming a net LNG exporter in 2016 and an overall net natural gas exporter in 2020. The price projections, which extend to 2040, never reach the annual average high reached in 2008 (see Figure 3).

Lower natural gas prices since 2008 along with a large and growing resource base have prompted calls for greater use of natural gas in the U.S. fuel mix. This is one of the key arguments against exporting U.S. natural gas. Natural gas comprised about 30% of U.S. primary energy consumption last year and has averaged 24% per year since 1973. Instead of exporting U.S. natural gas, some say, the United States should increase use of natural gas in the electric power sector to displace coal, as an alternative transportation fuel to displace oil, and to provide fuel and feedstock to domestic industries such as petrochemicals. Some transition to natural gas is already occurring, particularly in the electric power sector.[8]

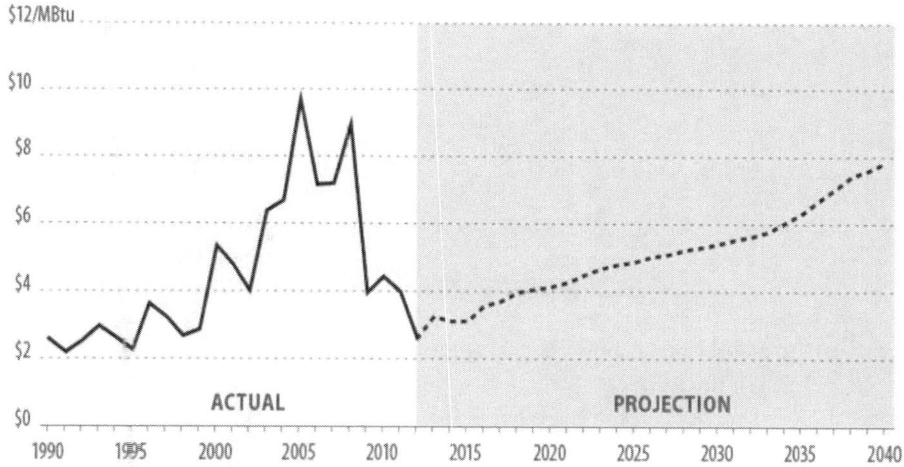

Source: EIA's natural gas price database, http://www.eia.gov/dnav/ng/hist/n9190 us3a.htm and EIA's Annual Energy Outlook 2013 Early Release, Natural Gas Section, reference case, http://www.eia.gov/oiaf/aeo/ tablebrowser/.

Notes: Prices reflect the average price at the wellhead in the lower 48. Projections are for 2011 forward. Units are dollars per thousand cubic feet ($/1,000 cu ft) in 2009 dollars.

Figure 3. Anual U.S. Natural Gas Prices, 1990-2040; Nominal dollars.

U.S. Natural Gas Exports to Date

Total U.S. natural gas exports are relatively small but have grown since 1999, increasing more than nine-fold through 2012. The United States has been exporting natural gas since at least the 1930s, primarily to Canada and Mexico.[9] In 2012, 98% of exports were by pipeline to Canada and Mexico. Starting in 1969, a small amount of natural gas was also exported as LNG via the Kenai LNG terminal in Nikiski, Alaska. Kenai LNG operated continuously from its opening until its idling in 2012, and remains the only LNG export facility in North America. Production of natural gas in the Cook Inlet of Alaska that supplies natural gas to Kenai LNG has declined too much to keep the facility operating.[10]

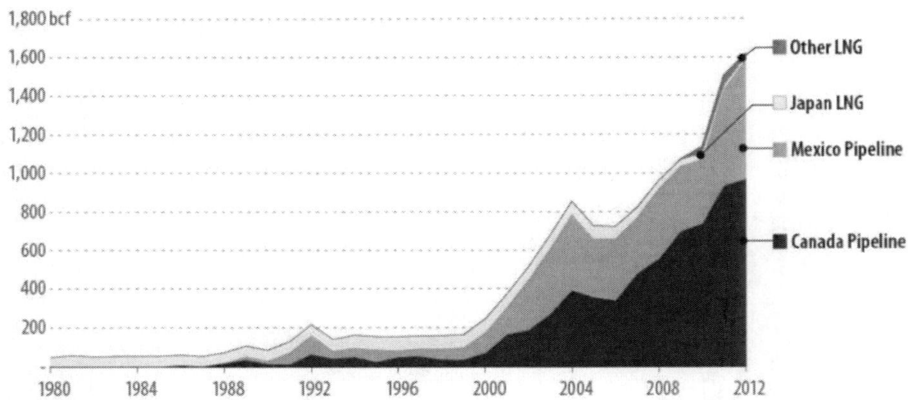

Source: EIA natural gas exports database, http://www.eia.doe.gov/dnav/ng/ng_move_expc_s1_a.htm.

Notes: LNG exports by volume are negligible to other countries besides Japan. Other LNG includes volumes that are re-exported from the United States after being imported from third countries. Units are billion cubic feet (bcf) per year.

Figure 4. U.S. Natural Gas Exports; 1980-2012.

In 2010, the United States, through DOE's Office of Fossil Energy, allowed LNG import terminals to receive cargos of LNG, store the LNG, and to re-export the LNG to international markets. Seven companies have received permits to re-export LNG with four pending. This explains the "Other LNG" category that emerged in 2010 (see Figure 4). Prior to 2010, LNG exports had declined steadily since 2005. Growing U.S. natural gas production, primarily from shale gas, has decreased the demand for both pipeline imports and LNG imports, leaving the import terminals mostly idle. Additionally, eight of the import terminals have applied for licenses to construct liquefaction facilities to export domestically produced LNG, while there are 23 applications for new liquefaction facilities, including floating facilities (see Table 1). In some cases companies that have partnered in a facility have both applied for licenses, but the volumes are only included once in DOE's data.

Pipeline Exports Increase

Natural gas exports by pipeline have risen since 1999, increasing over 16 times through 2012 and accounting for about 98% of total natural gas exports last year. Canada and Mexico, both free trade partners, are the only recipients of U.S. natural gas exports by pipeline. As countries with which the United States has free trade agreements (FTAs), exports of natural gas to both are assumed to be in the national interest by statute, thereby expediting the approval process for projects. The St. Clair, MI, transit point to Canada and the Roma, TX, transit point to Mexico are the busiest for U.S. natural gas exports (see Figure A-1).

Gross exports to Canada and Mexico have both increased since 1999, growing 25-fold and 10- fold, respectively. In 2012, Mexico's imports jumped almost 25% over the previous year. Facilitated by the 1994 North American Free Trade Agreement, new cross border natural gas pipelines have expedited the trade. Canadian natural gas production has declined almost 17% since peaking in 2002, but still remains above the level of Canadian consumption. Some of Canada's imports of natural gas from the United States are from gas produced in Canada's western provinces, imported into the United States, and re-exported to Canada's eastern provinces. This is a cost-effective way to transport the natural gas given pipeline constraints within Canada. Mexican natural gas consumption has increased about 80% since 2002, growing more than production.

LNG Activity on the Move

U.S. LNG exports started in 1969 with the opening of the Kenai LNG export terminal in Nikiski, Alaska. Japan, the world's largest LNG importer in 2011, received almost 100% of U.S. LNG exports until recent years. The project has not been expanded from its inception and remains the only U.S. LNG export facility. From 2005 to 2010, exports from Kenai LNG declined almost 50% due to depletion of its natural gas supply. The facility is being maintained, but no gas is flowing through it.

LNG Re-Exports: A Temporary Fix

Starting in 2010, DOE's Office of Fossil Energy, according to provisions in the Natural Gas Act, authorized LNG import terminals to receive LNG cargos from foreign countries and then reexport the LNG to other countries. Some LNG exporters try to take advantage of the idle U.S. import terminals and storage to wait for higher world prices. This trend almost doubled U.S. LNG exports to other countries, including new recipients Brazil, India, Spain, and the United Kingdom. Using the facilities for re-export helps maintain their operating capabilities in light of significantly decreased use to import LNG. Currently, seven companies have received permission to re-export LNG cargos from foreign countries with four applications pending.[11]

In order to re-export LNG, minimal or no additional equipment may need to be added to an import terminal. As mentioned above, DOE's Office of Fossil Energy must approve the change as does FERC.

Exports of Domestically Produced Natural Gas as LNG

There have been 31 applications for permits to export domestically produced natural gas as LNG, with a cumulative capacity of almost 12,344 billion cubic feet (bcf) per year or almost 50% of current U.S. production.[12] Eight of these liquefaction projects plan to adapt an existing LNG import terminal for export, which would require construction of liquefaction facilities at the import terminals, a major financial investment, with cost estimates ranging from $6 billion to $10 billion per terminal, mostly depending on capacity size. The other projects would construct new LNG export terminals, costing in the range of $20 billion each. Almost all the projects have received approval to export to free trade countries, but only the Sabine Pass Liquefaction (only its first application), Freeport LNG Expansion (only its first application), Lake Charles, and Dominion Cove Point projects have received DOE approval to export U.S. produced natural gas to non-free-trade countries.[13] Of the FTA countries, only Canada, Chile, Dominican Republic, Mexico, and South Korea have existing LNG import terminals. South Korea is the second-largest importer of LNG globally.

Table 1. Proposed North American LNG Export Projects

Company	Location	Volume (bcf/d)	Status		
			DOE FTA/non-FTA	FERC/ MARAD	Construction
Sabine Pass Liquefaction	Louisiana	2.2	Approved/Approved	Approved	Approved
Freeport LNG	Texas	1.4 plus	Approved/Approved	Pending	
Expansion and FLNG Liquefaction[a]		1.4	Approved/Pending		
Lake Charles Exports[b]	Louisiana	2.0	Approved/Approved	Pending	
Carib Energy[c]	Not Applicable	0.03 to FTA	Approved/Pending		
		0.01 to non-FTA			
Dominion Cove Point LNG	Maryland	1.0	Approved/Approved	Pending	
Jordan Cove Energy Project	Oregon	1.2 to FTA	Approved/Pending	Pending	
		0.8 to non-FTA			
Cameron LNG	Louisiana	1.7	Approved/Pending	Pending	
Gulf Coast LNG Export	Texas	2.8	Approved/Pending		
Gulf LNG Liquefaction	Mississippi	1.5	Approved/Pending		
LNG Development	Oregon	1.25	Approved/Pending	Pending	
Company (Oregon LNG)					
SB Power Solutionsc	Not Applicable	0.07	Approved/Not Applicable		
Southern LNG Company	Georgia	0.5	Approved/Pending	Pending	
Excelerate Liquefaction Solutions	Texas	1.38	Approved/Pending	Pending	
Golden Pass Products	Texas	2.6	Approved/Pending	Pending	
Cheniere Marketing	Texas	2.1	Approved/Pending	Pending	

Table 1. (Continued)

Company	Location	Volume (bcf/d)	Status		
			DOE FTA/non-FTA	FERC/ MARAD	Construct ion
Main Pass Energy Hubd	Offshore Louisiana	3.22	Approved/Not Applicable		
CE FLNG	Louisiana	1.07	Approved/Pending	Pending	
Waller LNG Services	Louisiana	0.16	Approved/Not Applicable		
Pangea LNG (North America) Holdings	Texas	1.09	Approved/Pending		
Magnolia LNG	Louisiana	0.54	Approved/Not Applicable	Pending	
Trunkline LNG Export[b]	Louisiana	2.0	Approved/Pending	Pending	
Gasfin Development USA	Louisiana	0.2	Approved/Not Applicable		
Freeport-McMoRan Energy[d]	Offshore Louisiana	3.22	Approved/Pending		
Sabine Pass Liquefaction[e]	Louisiana	0.28	Approved/Pending		
Sabine Pass Liquefaction[e]	Louisiana	0.28	Approved/Pending		
Venture Global LNG	Louisiana	0.67	Pending/Pending		
Advanced Energy Solutions[c]	Florida	0.02	Pending/Not Applicable		
Argent Marinec	NA	0.003	Pending/Not Applicable		
Eos LNG[d]	Texas	1.6	Pending/Pending		
Barca LNG[d]	Texas	1.6	Pending/Pending		

Source: Department of Energy, Office of Fossil Energy, Natural Gas Regulatory Responsibilities databases, http://fossil.energy.gov/programs/gasregulation/reports /summary_lng_applications.pdf.

Notes:

[a] DOE received a new application from FLEX to add 1.4 bcf/d of capacity at the Freeport LNG terminal for both FTA and non-FTA countries.

[b] Lake Charles Exports and Trunkline LNG Export have both filed to export LNG from the Lake Charles Terminal. The 2.0 bcf/d volumes are not additive.

[c] Project will use cryogenic containers to export LNG in small amounts on cargo ships and does not need a specialized export terminal like the other projects.

[d] Main Pass Energy Hub, LLC (LCE) and Freeport McMoRan Energy LLC (FME) have both filed to export from the Main Pass Energy Hub. The 3.22 bcf/d volumes are not additive and only FME includes exports to non-FTA countries. As an offshore facility project, these companies have applied to the Maritime Administration (MARAD), which improves offshore facilities instead of FERC.

[e] This is an expansion of the Sabine Pass Liquefaction project at the top of the table.

Adding liquefaction capacity will require new infrastructure to be added to the existing import terminals. The modification or expansion of existing facilities, including liquefaction trains, storage tanks, compressors, piping, and other equipment, will require authorization from FERC. Depending on the nature of individual facility modifications, compliance with additional state or federal statutory or regulatory requirements may also be required. For example, facility modifications would be subject to some level of environmental review pursuant to the National Environmental Policy Act (NEPA). Potential regulatory requirements are discussed below (see "Authorizations to Export LNG").

Natural gas in Alaska is also attracting attention as a possible source for exports. The cost of producing and transporting Alaskan natural gas to the lower 48 for consumption is high when compared with current U.S. prices. As Alaskan oil production declines, companies may seek to monetize or sell their natural gas, which is typically re-injected into oil wells to boost production. There are also new potential natural gas supplies in Alaska that may be more conducive to export compared to North Slope natural gas. Potential Alaskan projects have not been included in the DOE-sponsored studies on exports as Alaska is considered an independent market from the lower 48.

Trade: Agreements or Disagreements?

Most companies seeking permits to export LNG have applied to export LNG to countries with which the United States does not have a free trade agreement (FTA) in addition to those with which the United States does have an FTA. As mentioned above, exports to FTA countries are presumptively considered "in the national interest" under the Natural Gas Act, as amended. Five of those applications are pending with DOE. Of the 33.8 bcf/d of capacity applied for export by companies, 32.4 bcf/d or 96% is seeking or has received approval to export to non-FTA countries. As noted above, South Korea is the only major importer of LNG of the countries with which the United States has a free trade agreement. Of the other FTA countries, four have LNG import terminals, while the rest export natural gas, receive natural gas via pipeline, or do not import natural gas. In order for LNG export projects to be financially viable, they will likely need the ability to export to non-FTA countries.

The United States is in negotiations on two multi-nation free trade agreements: the Trans-Pacific Partnership (TPP) and the Transatlantic Trade and Investment Partnership (TTIP).[14] Both these agreements would give the signatories free trade status when it comes to U.S. natural gas exports. TPP includes Australia, Brunei, Canada, Chile, Japan, Malaysia, Mexico, New Zealand, Peru, Singapore, and Vietnam. Japan is the largest LNG importer in the world and would add significantly to potential markets for U.S. LNG exports. Of the four U.S. LNG export projects that have received approval to export to non-FTA countries, Japanese companies have signed contracts with two of the projects. Japanese companies also have signed contracts with at least two of the pending export projects. Japan was a late participant in TPP process; one of its motivations for joining was access to U.S. LNG exports.[15] TTIP is a proposed agreement between the United States and the European Union. Among other negotiating objectives, the EU may seek to gain access to U.S. LNG through the TTIP to lessen its dependence on Russian and other sources of natural gas.

The prospect of the United States limiting or restricting LNG exports has raised questions, particularly as a member of the World Trade Organization (WTO). The General Agreement on Tariffs and Trade's (GATT) Article XI, General Prohibition Against Quantitative Restraints, states:

> No prohibition or restrictions other than duties, taxes or other charges made effective through quotas, import or export licenses or other measures, shall be instituted or maintained by any contracting party on the importation of any product of the territory of any other contracting party or on the exportation or sale for export of any product destined for the territory of any other contracting party.

There are exceptions to Article XI based on the conservation of exhaustible natural resources or the necessity to protect human health, which may apply if the United States restricts LNG exports.[16] However, these exceptions may be dependent on a country restricting its own production. Additionally, restricting LNG exports may put the United States in a contradictory position vis-à-vis cases it has brought to the WTO, specifically against China for limiting the export of rare earths and other metals. The position of the United States as a promoter of free trade may also be challenged.

Natural gas imports and exports comprise a small fraction of overall U.S. international trade, totaling about $22 billion in 2012, with about two-thirds of the value coming from imports.[17] The United States was a net importer of natural gas of almost $8 billion. LNG imports and exports accounted for 13% of the trade. Since 2000, the value of exports has averaged almost $2.5 billion per year while imports averaged over $20 billion.[18]

THE GLOBAL LNG MARKET

If all the proposed U.S. LNG export projects were operational today, the United States would rank first in the world for global export capacity. However, U.S. LNG exports will face competition in the global LNG market. According to one study, global liquefaction capacity is projected to rise by almost 50% by 2020 (see Figure 5), including only one of the U.S. projects. Many non-U.S. projects are much further along than the U.S. projects. LNG trade slightly dropped year-on-year in 2012, accounting for 32% of all natural gas traded internationally.[19] Most LNG sold in the world is under long-term contracts, indexed to oil prices. The long-term contracts are needed to finance the liquefaction facilities, usually the most expensive part of the LNG supply chain, which includes LNG tankers, storage, and LNG import terminals. U.S. natural gas prices are market-based, which should give U.S. LNG export projects an advantage as the differential with oil-indexed-priced natural gas can be more than double the U.S. price (see Figure 2). U.S. LNG exports could add to the pressure for other countries to delink their natural gas exports—either as LNG or by pipeline—from oil-indexed prices. Japanese companies have been vocal about their interest in a natural gas-based pricing mechanism to reduce costs and exposure to oil prices. However, U.S. LNG export projects will still need to enter into long-term supply contracts, usually 20 to 30 years, to obtain financing, which may be a difficult hurdle to overcome given existing market and financial conditions. Providing LNG to countries that use oil for heating, industrial processes, or electricity generation could also decrease demand for petroleum products, putting downward pressure on oil prices.

The dip in 2012 in Figure 5 is due to major declines in Algeria, Trinidad & Tobago, Oman, and Indonesia. Algeria's two export terminals have been undergoing maintenance and one is being rebuilt. Trinidad & Tobago's decline is mostly because the United States is no longer a big importer of LNG. Egypt and Libya declined because of the turmoil in those countries.

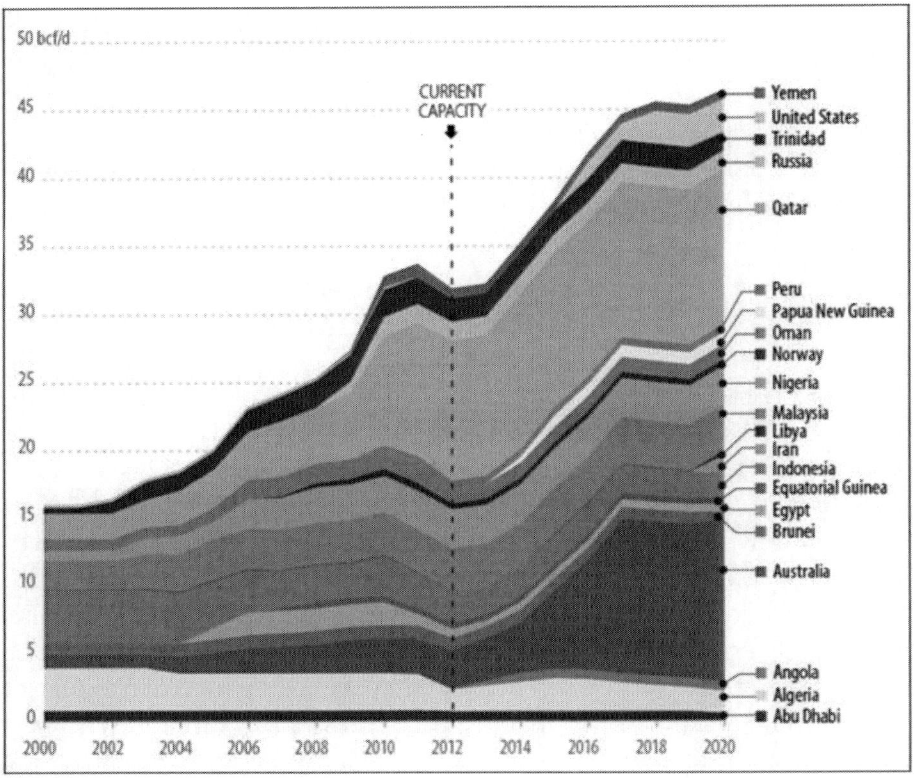

Source: PIRA Energy Group.

Notes: This graphic includes projects that are operating, under construction, or have reached a final investment decision. The only new U.S. project included is Cheniere's Sabine Pass terminal. The other proposed U.S. projects are considered speculative at this point. Units = billion cubic feet per day (bcf/d).

Figure 5. Actual and Projected LNG Production Capacity; 2000-2020.

Many of the projected projects in Figure 5 are targeting the Asian LNG demand centers. Although the locations of most of the proposed U.S. export terminals are on the U.S. Gulf Coast and the East Coast, Asia may be the target market for U.S. LNG as it tends to pay higher prices for its natural gas imports. The widening of the Panama Canal, scheduled to be completed in 2015, would contribute to U.S. competitiveness in Asia. Europe has a lot of LNG import capacity and growing demand, but needs to continue to improve its infrastructure connections to transport gas to markets.[20] Russia, the main supplier of natural gas to Europe, may be put under increasing pressure by U.S. export projects to further delink its natural gas prices from oil. U.S. LNG exports could also provide options for some countries that are highly dependent on one supplier.

AUTHORIZATIONS TO EXPORT LNG

Pursuant to provisions included under Section 3 of the NGA (15 U.S.C. §717b), both the export of LNG and the construction or expansion of LNG terminals require authorization from DOE's Office of Fossil Energy and from the Federal Energy Regulatory Commission. With regard to exports, any person seeking authorization to export LNG from the United States, or to amend an existing import or export authorization, must file an application with DOE's Office of Fossil Energy.[21] Denial of an authorization is dependent upon the export being deemed "not consistent with the public interest." That is, there is a presumption that exports to non-FTA countries are in the public interest unless shown otherwise.[22] If the United States has an FTA in effect with the nation to which the LNG would be exported, that application will be automatically deemed consistent with the public interest.[23] LNG exports to non-FTA countries may also be authorized, but require the Office of Fossil Energy to publish a notice of the application in the *Federal Register* and seek public comments, protests, and notices of intervention as part of their public interest determination. DOE also can limit the amount of cumulative LNG exports, so each successive project may be contingent upon the volumes of previously approved projects.

Also under the NGA, any person proposing to site, construct, or operate facilities to be used for the export of natural gas from the United States to a foreign country or to amend an existing FERC authorization, including the modification of existing authorized facilities, must file an application for authorization with FERC.[24] In addition to FERC, the Department of Homeland Security's U.S. Coast Guard and the Department of Transportation's Office of Pipeline Safety (OPS), under the Pipeline and Hazardous Materials Safety Administration (PHMSA), may also be responsible for exercising some level of regulatory authority over the siting, design, construction, expansion, and operation of LNG facilities, and related land and marine safety and security issues. Projects related to LNG exports or LNG terminals may be subject to requirements under other federal law, including the Coastal Zone Management Act, the Clean Air Act, and the Clean Water Act.

Compliance with one environmental law applies to all DOE authorizations required under Section 3 of the NGA—the National Environmental Policy Act

(NEPA, 42 U.S.C. 4321 *et seq.*). Pursuant to NEPA, DOE and other agencies are required to identify and document the environmental impacts of an action *before* a final agency decision is made. For projects related to LNG exports, the level of environmental review and documentation required will generally depend on whether the proposed action or project will require the construction of major new natural gas pipelines or related facilities or minor modifications of existing pipelines or related facilities.

DOE and the Public Interest Determination

As part of its process in determining whether LNG exports to non-FTA countries are "not consistent with the public interest," DOE's Office of Fossil Energy commissioned two studies: (1) a domestic price impact study by EIA and released in January 2012,[25] and (2) an economic impact study by NERA Economic Consulting (NERA) and released in December 2012. On February 25, 2013, DOE closed the comment period regarding the NERA study.[26] Almost 200,000 comments were received on the two DOE studies.[27] However, the results of these studies are only part of DOE's criteria for determining whether LNG exports to non-FTA countries are or are not in the public interest. In its approval of Cheniere Energy's non-FTA permit, DOE listed other criteria it used to make that determination: domestic need, adequacy of supply, the environment, geopolitics, and energy security. DOE has also provided insights into the "public interest" evaluation in a set of Policy Guidelines issued in 1984, Order No. 1471, and Delegation Order No. 0204-111. These were mostly to assess imports, but DOE has held that they also apply to exports.[28]

DOE has not laid out a specific timetable for when it will process the pending applications. In March 2013, the acting Assistant Secretary for Fossil Energy at DOE, Christopher Smith, declined to comment on a timetable for approving the non-FTA permits at a hearing before the House Oversight and Government Reform's Energy Policy, Health Care and Entitlements Subcommittee.[29] DOE has stated that it would begin "processing" the non-FTA applications according to:

1) All pending DOE applications where the applicant has received approval (either on or before December 5, 2012) from the FERC to use the FERC pre-filing process, in the order the DOE application was received,

2) Pending DOE applications in which the applicant did not receive approval (either on or before December 5, 2012) from FERC to use the FERC pre-filing process, in the order the DOE application was received,
3) Future DOE applications, in the order the DOE applications are received.[30]

The December 5, 2012, date coincides with when the NERA study was publicly released. The basic conclusion of the NERA study was that LNG exports would benefit the overall U.S. economy. The study acknowledged that there would be parts of the economy that would be hurt by LNG exports, mainly large consumers of natural gas, but that on a net basis the U.S. economy would be better off in all export cases. The net benefits would be highest if the United States could produce large quantities of low cost shale gas and that global demand for natural gas increases rapidly.

Both the NERA economic study and the EIA price analysis have been criticized. When EIA released its price analysis, the scenarios it examined per DOE's request were viewed as unrealistic. The high export scenario EIA examined was for 12 bcf/d, which is now well below the cumulative volumes for which companies have applied to DOE. According to EIA's analysis, the range of increases to domestic prices was 9.6% to 32.5%. Proponents of exports emphasized the former figure, while opponents focused on the latter. Both reports left enough latitude in their results for supporters and opponents of exports to promote their opinions.

The NERA report has attracted most of the attention in the 113th Congress, given the timing of its release. The NERA study, which used the EIA results, factored in international market parameters and macroeconomic impacts on the U.S. economy from increased LNG exports. In all the scenarios NERA examined, 63 in total, there were net economic gains to the United States, in spite of higher domestic natural gas prices, and the greater the exports the greater the benefits. The trade benefits from LNG exports outweighed the higher costs to the domestic economy. Criticism of the NERA report from some Members of Congress has focused on the narrowness of its results, the use of outdated data, and incomplete market information, among other things.[31] Although the criticism is open to debate, it is important to remember that the NERA study is just one part of DOE's analysis of determining which projects may be in the public interest. In its conditional approval to export to non-FTA countries for the Cove Point LNG terminal, DOE defended its data use from EIA's Annual Energy Outlook 2011 due to timing, but also noted that critics often cite the subsequent rise in demand projected in EIA's Annual Energy Outlook 2012 and the Early Release data from EIA's Annual Energy Outlook 2013, but neglect to highlight the corresponding rise in domestic production.[32] Similar to EIA, NERA was given a set of parameters to examine and asked for the impact on the overall U.S. economy. A full analysis of the NERA and EIA studies, as well as the myriad of other studies that have also been undertaken by other groups, is beyond the scope of this report.

Compliance with NEPA and Additional Environmental Laws

Both the Office of Fossil Energy's requirement to authorize LNG exports and FERC's obligation to authorize LNG terminal facility siting, construction, or expansion are considered federal actions subject to compliance with the National Environmental Policy Act. Broadly, NEPA requires federal agencies to identify and consider the environmental impacts of an action and to inform the public of those impacts before a final agency decision is made.[33]

To ensure environmental impacts are indeed considered, before an authorization required under Section 3 of the NGA can be issued, DOE agencies must conduct the appropriate NEPA review. The level of documentation and analysis required for a given NEPA review will vary depending on the anticipated level of impacts of that project. Regulations implementing NEPA, promulgated by the Council in Environmental Quality (CEQ),[34] require federal agencies to—

- **Prepare an environmental impact statement (EIS)**—a full-scale review of the potential environmental impacts. An EIS is required when it is known that a proposed action will significantly affect the quality of the environment.
- **Prepare an environmental assessment (EA)**—a concise public document prepared when the significance of a proposed action is unclear. If it is determined that project impacts will be significant, an EIS must be prepared. Otherwise, a Finding of No Significant Impact (FONSI) may be issued.
- **Determine whether a Categorical Exclusion (CE) may be applied**—a category of actions that, based on agency experience, normally have no cumulatively significant impacts and, hence, are categorically excluded from the requirement to prepare an EIS or EA.

DOE supplemented the CEQ regulations to specify NEPA compliance requirements applicable to actions undertaken by its agencies.[35] Generally, the level of NEPA review required for authorizations related to LNG exports will depend on whether granting a request for authorization will result in the construction of major new natural gas pipelines or related facilities (such as siting and constructing a new LNG terminal) or a significant expansion and modification of existing pipelines or related facilities.

To date, the Office of Fossil Energy has determined that applications for authorizations to export LNG may be processed as categorical exclusions.[36] To make that determination, applicants must demonstrate that the project, among other requirements, will not violate applicable statutory, regulatory, or permit requirements for environment, safety, and health, or similar requirements of DOE or Executive Orders or involve extraordinary circumstances that may affect the significance of the environmental effects of the proposal.[37]

In regulations implementing NEPA applicable to FERC, DOE has identified the authorizations required under Section 3 of the NGA for the siting, construction, and operation of new LNG export facilities as actions that normally will require an EIS.[38] However, under certain conditions, FERC may first prepare an EA to determine whether an EIS is necessary. For example, for an application to authorize the modification of an existing LNG import terminal to allow for LNG export, FERC may prepare an EA and determine that the required modification will not result in significant environmental impacts. Such a determination might be made if, for example, the modification would require adding liquefaction and related equipment, but remain within the terminal's existing footprint.

To prepare the appropriate NEPA documentation for an LNG export-related authorization, the Office of Fossil Energy or FERC would be required to identify any other compliance requirements applicable to the authorization.[39] Applicable requirements may involve some level of input, analysis, or approval from another federal agency, or possibly a tribal or state agency. Following are selected federal statutes, including agencies that may have some jurisdiction over those requirements that may apply to the construction of a new LNG export terminal or the expansion of an existing LNG import facility to include export operations:

- Clean Water Act—the U.S. Army Corps of Engineers, the Environmental Protection Agency (EPA), the U.S. Coast Guard, and state environmental protection agencies;
- Clean Air Act—EPA and state environmental protection agencies;

- Endangered Species Act—the Department of the Interior's U.S. Fish and Wildlife Service, the National Oceanic and Atmospheric Administration (NOAA) Fisheries Service, and state natural resource agencies;
- National Historic Preservation Act—the Advisory Council on Historic Preservation, the State Historic Preservation Officer, or Tribal Historic Preservation Officer; and
- Rivers and Harbors Act—the U.S. Coast Guard.

In addition to meeting appropriate environmental requirements, an authorized project may require compliance with additional safety or security-related requirements implemented by agencies including, but not limited to:

- The Department of Transportation's Office of Pipeline Safety (OPS) within the Pipeline and Hazardous Materials Safety Administration (PHMSA);
- National Fire Protection Association (NFPA); and
- Federal Emergency Management Agency (FEMA).

New Sources of Natural Gas: The Game Changer

The growth in U.S. natural gas resources, particularly from shale gas, and the projected continued growth are what make increased natural gas exports a possibility. U.S. natural gas reserves have climbed 72% since 2000 and 49% since 2005. These data include reductions for natural gas extracted during the period and so are net increases. In recent years, the increase in reserves is mostly attributed to development of shale gas, which has grown from 10% of U.S. natural gas reserves in 2007 to 32% in 2010. By comparison, conventional U.S. natural gas reserves declined between 2007 and 2008, and fell again in 2010. Though the decline was marginal, it highlights the importance of shale gas to future U.S. natural gas production. Many industry analysts expect shale gas reserves to continue to rise and make up a greater portion of U.S. natural gas reserves unless new restrictions are placed on the industry, such as rules related to hydraulic fracturing, power plant emissions, etc.

In 2012, the United States produced and consumed more natural gas than it ever has—24 trillion cubic feet (tcf) and 24 tcf, respectively—while paying some of the lowest market prices for natural gas in the world.[40] The production figure of 24 tcf is of dry gas, which has been processed for consumption purposes, but the United States actually produced 29.8 tcf of raw natural gas in 2012. The United States is the world's leading producer of natural gas, surpassing Russia in 2009, and the world's leading consumer. After declining for the first half of the last decade, U.S. natural gas production rose 18% in the latter half, with shale gas accounting for 30% of production by 2011.[41]

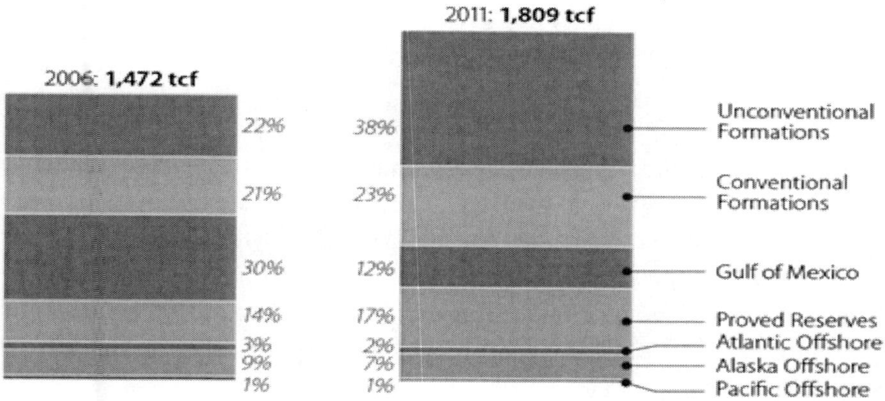

Source: Department of the Interior's U.S. Geological Survey and Bureau of Ocean Energy Management (BOEM) and U.S. Energy Information Administration.

Notes: The data for unconventional include some but not all of the shale basins, as some have not been assessed to date by the U.S. Geological Survey. Unconventional shales are fine grained, organic rich sedimentary rocks. The shales are both the source of and the reservoir for oil and natural gas, unlike conventional petroleum reservoirs. Undiscovered technically recoverable resources (UTRR) refers to amounts of natural gas estimated to exist by examining geologic characteristics of unexplored areas and recoverable using current technology. All the figures in the graphic above are UTRR except Proved Reserves, defined as a 90% probability of recovery using existing technology and at current prices. Units = trillion cubic feet (tcf).

Figure 6. Natural Gas Resources and Reserves; 2006 vs. 2011.

Reserves and production data do not tell the whole story when looking at the U.S. transformation regarding natural gas supply. The term *reserves* has a specific industry definition that includes a technological component, an economic factor, and a probability of success among other criteria. To more fully understand the changes to the U.S. natural gas sector it is more appropriate to look at reserves and estimates for undiscovered, technically recoverable resources (UTRR) (see Figure 6). UTRR is an estimate of what can be extracted using current technology regardless of price. Using UTRR plus reserves, the United States has a natural gas resource base of 1,809 tcf or enough gas for approximately 79 years of production at 2011 levels. Compared with data from 2006, U.S. UTRR for natural gas has jumped almost 25%. Even this measure may not accurately reflect what will be extracted from the ground as technology is constantly changing. Just over the last few years, industry has been able to improve its shale gas extraction rate from about 5% to about 15%, thereby tripling what is recoverable.

Projected Future Growth

In 2012, natural gas was the most produced fuel, on an energy equivalent basis, in the United States, surpassing coal for the first time. This change was driven by the success of shale gas development. EIA, which makes projections based on current policy and information, estimated in its *Annual Energy Outlook 2013 Early Release* that overall U.S. natural gas production will grow 55% between 2010 and 2040. Shale gas will comprise over 50% of that production, up from 23% in 2010.

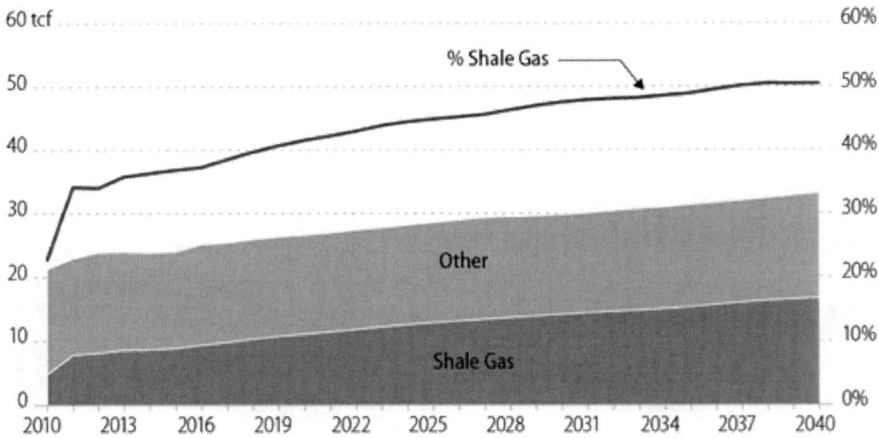

Source: EIA Annual Energy Outlook 2013 Early Release, Reference Case, http://www.eia.gov/oiaf/aeo/ tablebrowser/#release=AEO2013ER&subject=8-AEO2013ER&table=14-AEO2013ER&region=0-0&cases= early2013-d102312a.

Note: Units = trillion cubic feet (tcf) per year.

Figure 7. Projected U.S. Natural Gas Production.

During that time period, the United States is expected to go from a net importer of natural gas by pipeline and LNG to a net exporter by 2020, which is a change from EIA's *2011 Annual Energy Outlook* when there was no time period in which the United States was forecast to be a net exporter of natural gas. The United States is forecast to be a net LNG exporter by 2016, according to EIA.

Natural Gas Liquids: A Production Driver

Natural gas liquids (NGLs) have taken on a new prominence as shale gas production has increased and prices have fallen. NGL is a general term for all liquid products separated from natural gas at a gas processing plant and includes ethane, propane, butane, and pentanes. When NGLs are present with methane, which is the primary component of natural gas, the natural gas is referred to as either "hot" or "wet" gas. Once the NGLs are removed from the methane the natural gas is referred to as "dry" gas, which is what most consumers use. Each NGL has its own market and its own value. As the price for dry gas has dropped because of the increase in supply and other reasons such as the warm winter of 2011, the natural gas industry has turned its attention to producing more wet gas in order to bolster the value it receives. Some companies have shifted their production portfolios to tight oil formations, such as the Bakken in North Dakota, to capitalize on the experience they gained in shale gas development. Historically, the individual NGL products have been priced against oil, except for ethane, and as oil prices have remained higher since 2005 relative to natural gas, it has driven an increase of wet gas production, thereby maintaining the amount of dry gas as a production "byproduct" despite its low price.

CONGRESSIONAL ACTIONS AND CONSIDERATIONS

There have been four bills introduced in the 113th Congress directly related to LNG exports, primarily to expand the list of countries that would get automatic approval beyond FTA countries. Other bills have been introduced that affect natural gas fundaments and infrastructure.

The Senate Energy and Natural Resources Committee conducted a hearing on natural gas, including exports, on February 12, 2013, and the House Oversight and Government Reform's Energy Policy, Health Care and Entitlements Subcommittee held a hearing on the DOE's permitting process on March 19, 2013. There may be additional hearings in the coming months.

Some congressional representatives from the states where the LNG facilities are planned have submitted letters of support for the different projects to various regulatory agencies. Additionally, other Members have mentioned the possibility of natural gas exports in public statements, and some of those have expressed concerns about the potential effect on domestic prices, supplies, and the environment, among other issues.

Issues and Interests

The public focus on U.S. exports of natural gas has been on the applications to export LNG, despite the United States exporting much more natural gas by pipeline. Groups such as the Industrial Energy Consumers of America (IECA), a national association of manufacturing companies, and the American Public Gas Association (APGA), a national association of publicly-owned natural gas distribution systems, have filed motions to intervene against various projects.[42] Both of these organizations represent firms that use natural gas and would be negatively affected if natural gas prices rose. Natural gas producers and certain local businesses have supported the projects as they would benefit from access to new overseas markets and higher international prices. Analyses of the price effects of potential natural gas exports likely will receive continued attention as companies move forward with their projects. The Sierra Club has filed against projects on environmental grounds, particularly related to the source of natural gas for export.

Expectations about the economic impacts of greater U.S. natural gas exports depend on assumptions about the volume of exports, economic growth, market segmentation, and environmental regulations, among other market parameters. Some initial estimates projected a modest rise in absolute terms in domestic natural gas prices if all the proposed export projects are built, premised on a relatively flat supply curve for natural gas. These estimates also projected that U.S. natural gas prices would stay relatively low in historic terms as well as in comparison to global prices. Those in favor of exports add that increased production could result in increased revenue for local, state, and federal governments (through taxes, royalty payments, and economic development), more employment, an improved international trade balance, and reductions in natural gas flaring.[43] Natural gas consumers counter that higher natural gas prices abroad could eventually lead to higher prices in the United States, and possible supply shortages, as producers seek to maximize profits by diverting more and more U.S. natural gas to overseas markets.[44]

In the near term, increased use of natural gas in the U.S. economy is limited, primarily to electric power generation.[45] Natural gas-fired electric power plants account for a significant and growing share of U.S. natural gas demand. Although coal remains the dominant fuel for U.S. electric power generation, environmental concerns regarding atmospheric emissions is limiting its use and prompting the retirement of older coal plants that are less equipped to curtail emissions. Switching from coal to natural gas in electric power generation may consume incremental U.S. natural gas supply increases before exports do. There are many proposed petrochemicals projects, but these are at various stages of development and will take a number of years to come to fruition. Other sectors such as transportation, industrial, commercial, and residential are not likely to see a substantial rise in natural gas demand in the next couple of years. This could change if technologies can be improved to increase the use of natural gas in transportation, such as gas-toliquids, natural gas vehicles, or electric vehicles (assuming the electricity is generated by natural gas). Although proponents see strategic value in such fuel-switching as a means to reduce U.S. dependence on imported oil, high technology costs diminish this prospect in the near term.

Although much less attention is paid to natural gas exports by pipeline, it is anticipated that these will continue to increase as more shale gas is developed.

Canada's natural gas production has been declining, but it is possible this will be reversed as Canada develops its own shale gas resources, which are estimated to be large. However, Canadian consumption may also increase as production from oil sands is projected to rise. Natural gas is heavily used in the extraction of petroleum from oil sands. Canada also has at least two of its own LNG export projects being considered. A recent EIA study estimated Canada's technically recoverable shale gas resources at 573 trillion cubic feet, over 100 years' worth at the country's current production rate.[46] If Canada develops these resources, they could be an additional source of natural gas for the United States as well.

Mexico's natural gas production has been rising steadily for the last decade, but not quickly enough to keep up with consumption. Imports now account for almost 27% of consumption compared to under 10% in 2000, and imports from the United States made up over three-quarters of all natural gas imports to Mexico in 2012.[47]

Although Mexico may have significant technically recoverable shale gas resources , 545 trillion cubic feet or 364 years at their current production level, Mexico is much further behind in developing these resources than the United States and Canada,[48] and will likely remain dependent on U.S. supplies to meet growing demand.

Receiving permits to export natural gas by pipeline to Canada and Mexico is typically easier than receiving a permit to build an LNG export facility, even though pipeline projects require authorization from the Secretaries of Defense and State. Both Canada and Mexico are FTA countries and exports are assumed to be in the U.S. national interest by statute. Pipeline export projects tend to be less costly and easier to finance than LNG export projects may be; none of the latter have been built in the United States in 40 years.

As highlighted above, the development of shale gas resources will be a key factor in the United States becoming a net natural gas exporter. Infrastructure constraints within some of the major shale gas producing areas may limit the amount of natural gas that can reach markets and be available for export. Changes to the regulatory environment would also have an impact on natural gas production.

Environmental groups differ on the desirability of greater natural gas use in general. Although burning natural gas produces less pollution than burning other fossil fuels, it still emits greenhouse gases and other atmospheric pollutants. Some environmental groups view natural gas as a necessary bridge fuel to a zero carbon economy, while others want to go to the zero carbon economy directly.

Some environmental groups see natural gas exports raising domestic natural gas prices, making renewables more viable. Additionally, there are concerns about risks to water supplies associated with hydraulic fracturing, the technique for extracting shale gas which uses water, sand, and chemicals to create fissures in shale, allowing the trapped natural gas to be cost-effectively extracted. The possibility of increased shale gas development and pipeline construction in the United States to supply overseas LNG buyers troubles some environmental advocates.

With natural gas prices low and projected to remain so, producers want new markets for their product. Exports represent one alternative outlet for natural gas.

APPENDIX A. SELECT U.S. NATURAL GAS IMPORT AND EXPORT INFRASTRUCTURE

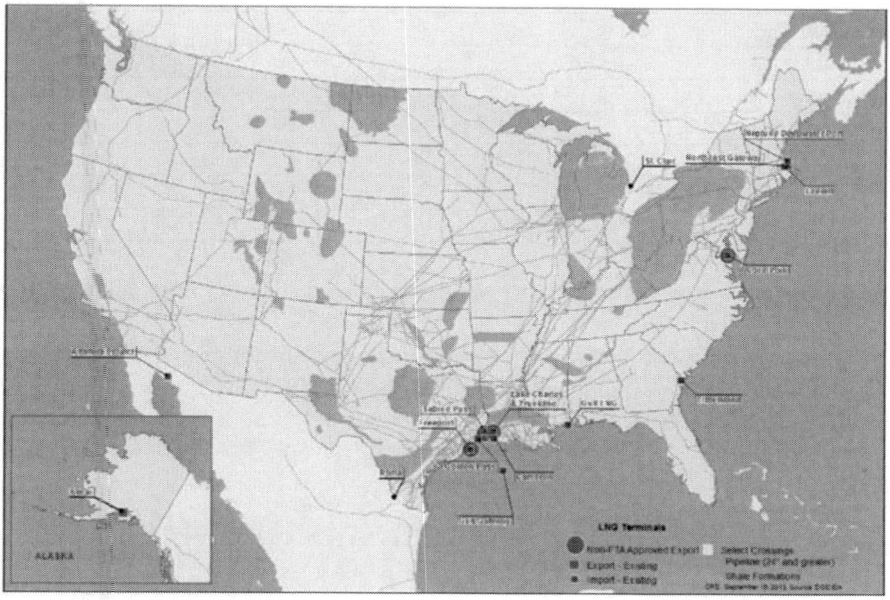

Source: Compiled by CRS from EIA sources.

Notes: Hawaii is not shown on this map because it has very limited natural gas infrastructure.

Figure A-1. Select U.S. Natural Gas Import and Export Infrastructure.

APPENDIX B. SUPPLY/DEMAND BALANCE

Table B-1. 2012 U.S. Supply and Demand Balances

	Volume (bcf)	Percentage of Consumption
Consumption	**25,457**	**100%**
Commercial	2,905	11%
Electric Power	9,137	36%
Industrial	7,100	28%
Residential	4,177	16%
Other	2,138	8%
Production[a]	**24,048**	**94%**
Net Imports	**1,516**	**6%**

Table B-1. (Continued)

	Volume (bcf)	Percentage of Consumption
Imports	3,135	12%
- Canada	2,960	12%
- LNG	175	1%
Exports	1,619	6%
- Canada	971	4%
- LNG (Alaska)	14	<1%
- Re-Export	14	<1%
Change in Inventories	107	**<1%**

Source: U.S. Energy Information Administration databases, http://www.eia.gov /naturalgas/data.cfm

Notes: Units = billion cubic feet (bcf). Table shows data for major sub-categories, which may not equal the section total, and sums may not total due to rounding.

[a] Dry natural gas production.

APPENDIX C. NATURAL GAS HUB MAP

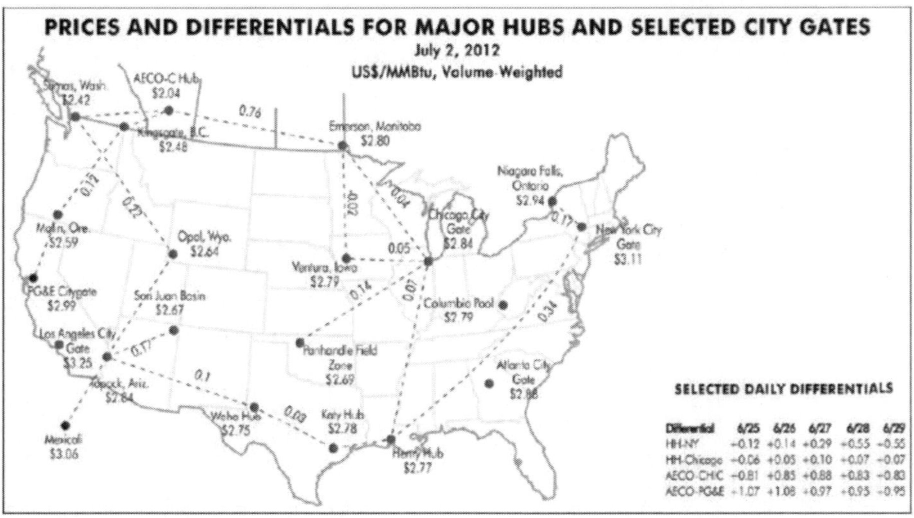

Source: Natural Gas Week, July 2, 2012.

Notes: The figures along the dotted lines indicate transportation costs between hubs and city gates. Alaska and Hawaii are not included on this map as their markets are distinct from the lower 48 states.

Figure C-1. Natural Gas Hub and City Gate Prices.

End Notes

[1] CRS held a seminar on LNG exports on March 1, 2013, for congressional staff. The seminar video is available for staff on the CRS website, http://www.crs.gov/programs/Pages /RecordedEventDetail.aspx?PRODCODE=WRE00058.

[2] Exports of natural gas from Alaska are viewed differently as those resources are isolated from the rest of the U.S. market because natural gas prices in the lower 48 do not justify building the required infrastructure to transport natural gas via pipeline or as LNG. Additionally, there may not be any, Jones Act-compliant LNG tankers to transport natural gas to the market in the lower 48 states. In DOE's analysis of U.S. LNG exports, it has not factored LNG exports from Alaska into its analysis regarding impacts on the U.S. economy and domestic prices.

[3] For additional information and analysis of shale gas and fracking see CRS Report R41760, *Hydraulic Fracturing and Safe Drinking Water Act Regulatory Issues*, by Mary Tiemann and Adam Vann.

[4] For additional information about natural gas and the U.S. economy see CRS Report R42814, *Natural Gas in the U.S. Economy: Opportunities for Growth* , by Robert Pirog and Michael Ratner

[5] U.S. Energy Information Administration (EIA), *Annual Energy Outlook 1999 with Projections to 2020*, DOE/EIA0383(99), Washington, DC, December 1998, p. 71, http://www.eia.gov/oiaf/archive/aeo99/pdf/0383(99).pdf.

[6] U.S. Energy Information Administration, *Annual Energy Outlook 2013 Early Release*, DOE/EIA-0383ER(2013), Washington, DC, December 5, 2012, http://www.eia.gov/oiaf /aeo/tablebrowser/#release=AEO2013ER&subject=8-AEO2013ER&table=13-AEO2013ER &region=0-0&cases=early2013-d102312a.

[7] For additional information on the U.S.-Canada energy relationship see CRS Report R41875, *The U.S.-Canada Energy Relationship: Joined at the Well*, by Paul W. Parfomak and Michael Ratner.

[8] For additional information about natural gas in electrical power generation, see CRS Report R42950, *Prospects for Coal in Electric Power and Industry*, by Richard J. Campbell, Peter Folger, and Phillip Brown.

[9] Data are unavailable earlier than the 1930s and distinguished by country since the 1950s.

[10] The facility has not been dismantled, but is being maintained. It is possible that the facility could reopen if Alaskan natural gas production increases in the future.

[11] Department of Energy, Office of Fossil Energy, *Status of Short-Term Applications to Export Previously Imported LNG*, March 5, 2013, http://fossil.energy.gov/programs/gas regulation/reports/Previously_Imported_LNG.pdf.

[12] DOE, Office of Fossil Energy, *Authorizations Database*, Washington, DC, January 30, 2013, http://fossil.energy.gov/ programs/gasregulation/reports/summary_lng_applications.pdf.

[13] Free Trade Agreement countries that require national treatment include Australia, Bahrain, Canada, Chile, Dominican Republic, El Salvador, Guatemala, Honduras, Jordan, Mexico, Morocco Nicaragua, Oman, Peru, Singapore, and South Korea.

[14] For additional information on the two agreements see CRS Report, *The Trans-Pacific Partnership Negotiations and Issues for Congress*, by Ian F. Fergusson and CRS Report R43158, *Proposed Transatlantic Trade and Investment Partnership (TTIP): In Brief*, by Shayerah Ilias Akhtar and Vivian C. Jones.

[15] Discussion with Amb. Koji Tsuruoka, Chief Negotiator for the Trans-Pacific Strategic Economic Partnership, August 6, 2013.

[16] See GATT Article XX.

[17] U.S. Energy Information Administration, *U.S. Imports by Point of Entry and U.S. Exports by Point of Exit*, Washington, DC, January 31, 2013, http://www.eia.gov/naturalgas /data.cfm#imports.

[18] Bureau of Economic Analysis, Department of Commerce, *Trade in Goods databases*, June 14, 2010, http://www.bea.gov/agency/uguide1.htm#_1_19.

[19] *BP Statistical Review of World Energy 2013*, June 2013, p. 29.

[20] For more information on Europe's natural gas situation, see CRS Report R42405, *Europe's Energy Security: Options and Challenges to Natural Gas Supply Diversification*, coordinated by Michael Ratner.

[21] Requirements applicable to LNG exports are specified under Section 3 of the NGA (15 U.S.C. §717b). Regulations implementing requirements applicable to the export authorization application process were established under 10 C.F.R. Part 590, the "Administrative Procedures with Respect to the Import and Export of Natural Gas."

[22] Applicants seeking authorization to export LNG may seek either a blanket or a long-term authorization. The blanket authorization enables the applicant to export on a short-term or spot market basis for up to two years. The long-term authorization is used when an applicant has, or intends to have, a signed gas purchase or sales agreement/contract for a period of time longer than two years.

[23] See 15 U.S.C.§717b(c). Regulations implementing this section of the NGA were promulgated under 18 C.F.R. Part 153, "Applications for Authorization to Construct, Operate, or Modify Facilities Used for the Export or Import of Natural Gas."

[24] Pursuant to 15 U.S.C. §717b(e)(1).

[25] For the complete EIA report follow this link, http://www.fossil.energy.gov /programs/gasregulation/LNGStudy.html.

[26] For the complete NERA report follow this link, http://www.fossil.energy.gov /programs/gasregulation/reports/ nera_lng_report.pdf.

[27] E-mail from DOE's Office of Fossil Energy, February 27, 2013.

[28] Department of Energy, Office of Fossil Energy, *Opinion and Order Conditionally Granting Long-Term Authorization to Export Liquefied Natural Gas from Sabine Pass LNG Terminal to Non-Free Trade Agreement Nations*, FE Docket No. 10-111-LNG, Washington, DC, May 20, 2011, pp. 28-29.

[29] U.S. Congress, House Oversight and Government Reform, Energy Policy, Health Care and Entitlements, *The Department of Energy's Strategy for Exporting Liquefied Natural Gas*, 113th Cong., 1st sess., March 19, 2013.

[30] Office of Oil Gas Global Security Supply, *Pending Long-Term Applications to Export LNG to Non-FTA Countries - Listed in Order DOE Will Commence Processing*, Office of Fossil Energy, U.S. Department of Energy, Washington, DC, September 10, 2013.

[31] Office of Senator Ron Wyden, "Wyden Highlights Flaws in DOE Export Study," press release, January 10, 2013, http://www.wyden.senate.gov/news/press-releases/wyden-highlights-flaws-in-doe-export-study-.

[32] U.S. Department of Energy, Office of Fossil Energy, Order Conditionally Granting Long-Term Multi-Contract Authorization to Export Liquefied Natural Gas by Vessel from the Cove Point LNG Terminal to Non-Free Trade Agreement Nations, DOE/FE Order No. 3331, Washington, DC, September 11, 2012, pp. 86-87, http://energy.gov/sites/prod/files /2013/09/f2/Order%203331.pdf.

[33] Under Section 102 of NEPA, all federal agencies are required to include in "every recommendation or report on proposals for legislation and other major Federal actions

significantly affecting the quality of the human environment, a detailed statement by the responsible official on—the environmental impact of the proposed action" (42 U.S.C. §4332(2)(C)(i)).

[34] CEQ regulations, at 40 C.F.R. §§1500-1508, are broadly applicable to all federal agency actions.

[35] DOE promulgated "National Environmental Policy Act Implementing Procedures" at 10 C.F.R. Part 1021. Those procedures apply to all organizational elements of DOE, except FERC. "Regulations Implementing the National Environmental Policy Act," at 18 C.F.R. Part 380, apply specifically to FERC actions.

[36] See the U.S. Department of Energy's Office of Fossil Energy webpage "Natural Gas Import & Export Regulation, NEPA: Records of Categorical Exclusions," at http://fossil.energy.gov /programs/gasregulation/ nepa_cx_determination.html.

[37] See the Office of Fossil Energy "Categorical Determination Form," issued February 6, 2013, for ENI USA Gas Marketing, LLC, application seeking authorization to export previously imported LNG from the Cameron LNG terminal in Cameron Parish, Louisiana, available at http://fossil.energy.gov/programs/gasregulation/cx_documents/ Cat_Ex_12_161_LNG.pdf.

[38] See 18 C.F.R. §380.6(a)(1).

[39] For additional information on siting a liquefied natural gas terminal see CRS Report RL32205, *Liquefied Natural Gas (LNG) Import Terminals: Siting, Safety, and Regulation*, by Paul W. Parfomak and Adam Vann.

[40] Many producing countries subsidize natural gas consumption, so their consumers do not pay a market price.

[41] 2011 is the latest year for which data are available for this metric.

[42] Letter from Paul N. Cicio, President of Industrial Energy Consumers of America, to U.S. Department of Energy, Office of Fossil Energy, Office of Oil and Gas Global Security and Supply December 13, 2011, http://www.iecaus.com/documents/121310.pdf.

American Public Gas Association, "APGA Files Motion to Intervene and Protest Freeport Export Application," press release, March 28, 2011, http://www.apga.org/files/public /Press%20Releases/2011/Press%20Release%20-%20Motion%20to%20Intervene%20in %20Freeport%20Application,%202011.pdf.

American Public Gas Association, "APGA Files Motion to Intervene in Sabine Pass LNG Export Facility Application," press release, March 4, 2011, http://www.apga.org/files/public /Press%20Releases/2011/Press%20Release%20-%20Comments%20Sabine%20Pass- %20March%203,%202011.pdf.

[43] Flaring is combusting natural gas as a means to eliminate it because it may be impracticable to use, capture, or transport. Flaring is usually done as a safety or health precaution, during the exploration and development phases leading to production.

[44] Margaret Ryan, "USAEE Notebook: DOE Weighing LNG Export Price Effect," *AOL Energy*, Internet blog, October 10, 2011, http://energy.aol.com/2011/10/10/usaee-notebook-doe-weighing-lng-export-price-effect/.

In: United States Fossil Fuel Exports
Editor: Mitchell P. Henning

ISBN: 978-1-63117-198-7

Chapter 3

U.S. Coal Exports*

Marc Humphries, John Frittelli, Jane A. Leggett, Linda Luther and Michael Ratner

Summary

The gap between available U.S. coal supply and demand may continue to widen as low cost natural gas becomes more attractive to electric power plants and uncertainties with emission regulations may inhibit new coal plant investments. Coal producers with excess supply will likely seek to expand their market abroad. Consequently, U.S. coal exports are forecast to continue to rise over the next decade and possibly longer. Growth potential in Asian markets seems large, but there are potential bottlenecks, such as infrastructure and potential rising costs of regulation, competition from other fuels, and transportation constraints that could slow export growth.

The U.S. Energy Information Administration (EIA), in its 2013 Annual Energy Outlook Early Release reference case, projects net exports of coal to trend up through 2040, almost 50% from 2011 levels, with some fluctuations. The significance of this may have short- and long-term ramifications as well as positive and negative consequences. Increased net exports could improve the U.S. trade balance as well as add government revenue from production that may otherwise decline because of falling domestic consumption. Environmentally, exporting coal may

* This is an edited, reformatted and augmented version of Congressional Research Service publication, No. R43198, dated August 21, 2013.

run counter to the current Administration's domestic environmental policies and affect U.S. efforts to address global environmental issues. Depending upon the nature of the coal exports, certain parts of the country may benefit economically. Current and projected coal exports, the associated infrastructure, and the environmental consequences have prompted interest by Congress.

The United States has been exporting coal since the late 1800s. From 2003 to 2012, U.S. coal exports have risen over 200%, mainly driven by competitive production costs, global demand, and lower prices, among other factors. Coal exports comprised 12% of U.S. coal production in 2012. In 2011, U.S. coal exports broke 100 million short tons (MST) for the first time since 1992 and in 2012 surpassed their peak of almost 113 MST in 1981. The value of U.S. coal exports has increased, rising from under $10 billion in 2010 to almost $16 billion in 2011, according to U.S. Energy Information Administration data.

Many factors will influence how much coal will be exported from the United States. Enough projects have been proposed that by 2016 the United States could be exporting more than double its current coal exports. Projects in the Pacific Northwest have attracted much of the attention, even though the Northeast continues to be the source for most exports.

Representatives from state and local agencies, particularly in Washington and Oregon, as well as industry, community, and environmental groups, along the potential coal transport corridor, have expressed outright support for or opposition to port terminal projects that would allow for increased export of Powder River Basin coal through the Pacific Northwest. Opponents have argued that increased train and barge traffic will have significant adverse impacts to the human, natural, and cultural environment. Project supporters have argued that the projects would create or maintain jobs in the construction, mining, and transportation industries and bring increased tax revenue to the states.

Broadly, the National Environmental Policy Act (NEPA) requires federal agencies to consider the environmental impacts of their actions before a final decision is made regarding that action.

INTRODUCTION

The current gap between U.S. coal supply and domestic demand may widen as low cost natural gas becomes more attractive to electric power plants, and uncertainties with new environmental regulations discourage investment in new coal plants. Coal producers with excess supply will likely seek to expand their market abroad. Consequently, in the long run (over the next decade and possibly longer), U.S. coal exports are expected to rise. In the near term, however, the U.S. Energy Information Administration (EIA) projects coal exports to decline from a high point of 126 million short tons in 2012 to 107 million short tons in 2014. Growth potential in Asian coal markets seems large, but there are potential bottlenecks (e.g., lack of infrastructure, potential rising costs of regulation, competition from other fuels, and transportation), which could slow export growth. Also, limits on natural gas supply (for any number of reasons) would likely increase domestic coal prices and cut into exports. Although U.S. consumption of coal appears to be declining, other parts of world are increasing their use of coal.

Table 1. U.S. Coal Supply-Demand Balance

	2011	**2012**
Total Supply	**1,121.9**	**1,040.9**
Production	1,095.6	1,020.5
Imports	13.1	9.2
Waste Coal Supplied	13.2	11.2
Total Demand	**1,110.2**	**1,015.0**
Consumption	1,002.9	889.3
Electric Power	932.5	824.8
Exports	107.3	125.7
Metallurgical	69.5	NA
Thermal	37.7	NA
Stock Change	0.2	8.2
Unaccounted	11.5	17.6
Exports/Production	9.8%	12.3%

Source: U.S. Energy Information Administration, Monthly Energy Review, March 27, 2013, http://www.eia.gov/ totalenergy/data/monthly/pdf/sec6.pdf.

Notes: Units = million short tons (MST). May not sum due to rounding. According to EIA, waste coal is usable material that is a byproduct of previous coal processing operations, usually composed of mixed coal, soil, and rock (mine waste). Most waste coal is burned as-is in unconventional fluidized-bed combustors. For some uses, waste coal may be partially cleaned by removing some extraneous noncombustible constituents. Examples of waste coal include fine coal, coal obtained from a refuse bank or slurry dam, anthracite culm, bituminous gob, and lignite waste.

U.S. consumption of coal declined from 2011 to 2012. Coal production also declined, but exports rose 17%.[1] The decrease in supply was driven, in part, by the decline in domestic demand. The EIA includes coal in its 2013

Annual Energy Outlook Early Release. In its reference case, net exports of coal may increase through 2040, increasing almost 50% from 2011 levels.[2] Increased net exports could improve the U.S. trade balance as well as add government revenue from production on federal lands that might otherwise decline because of falling domestic consumption. Exporting coal may run counter to the current Administration's domestic environmental policies aimed at reducing greenhouse gas emissions (GHG), and affect current U.S. policy goals addressing global climate change and other environmental issues. Depending upon the location of the coal exports, certain parts of the country may benefit economically.

Current and projected coal exports, issues associated with infrastructure, and the environment have prompted interest by Congress. Bills were introduced in the 112th Congress that supported increased coal exports (H.R. 3409, S. 3450) and sought to limit exports (H.R. 6202).

Coal export-related issues that have raised congressional concern include carbon emissions abroad, coal dust from rail transport, infrastructure development (particularly port expansion), and federal coal valuation for royalty purposes, among others.

WORLD COAL SUPPLY

Global production, consumption, and trading of coal have increased over the last decade, and are projected to continue growing. In fact, worldwide coal was the fastest-growing source of primary energy in 2011, according to the International Energy Agency's (IEA's) *Coal Medium-Term Market Report 2012*.[3] Coal consumption is projected to also grow more than oil or natural gas over the next five years, according to the IEA report, and reach parity with oil as the most used fuel source in the world.[4]

World Coal Production

Worldwide, coal production has increased by nearly 60% since 2002, with most of the increase coming from China—a 130% rise. China accounted for about 50% of coal production in 2011, up from 34% in 2002. U.S. coal production has declined by 13 million tonnes of oil equivalent (Mtoe) over this time period.[5] The data in Figure 1 illustrate that other countries such as Colombia, Indonesia, and India also had significant production increases since

2002. India's coal production grew by 60% over the past 10 years, while Indonesia's production more than tripled. Australia increased coal production by 26% over the same time period.

At the company level, the concentration of production worldwide is not significant. In 2010, the top five world coal producers accounted for about 18% of world production (Coal India (6%), Shenhua Group China (5%), Peabody Energy (3%), Datong Coal Mining Group-China (2%), and Arch Coal (2%)). The top 30 coal firms produced 40% of world production.[6]

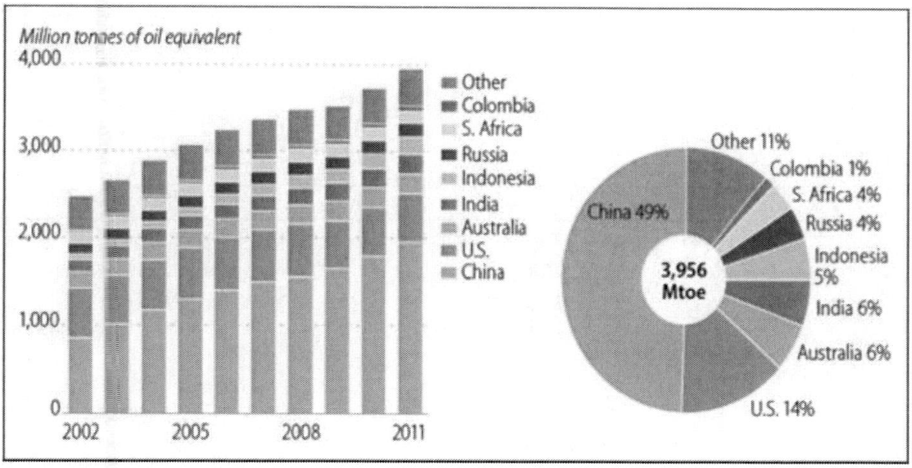

Source: BP Statistical Review of World Energy, 2012.
Notes: Pie chart based on 2011 data.

Figure 1.World Coal Production, 2002-2011.

World Coal Reserves and Resources

The United States ranked number one in the world in coal reserves in 2011, with almost 240 billion metric tons (mt) (28%).[7] Russia was second with an estimated reserve base of 157 billion mt (18%), with China in third with 115 billion mt (13%).

Taken together, the top three countries hold 59% of the world's recoverable coal reserves. When India and Australia are added, the top five coal producing countries hold 75% of world recoverable coal reserves (see Table 2).

Table 2. Global Proved Coal Reserves, 2011
Billion tones

Country	Bituminous and Anthracite	Sub-bituminous and Lignite	Total	Percent of Total
United States	108.5	128.8	237.3	28
Russia	49.1	107.9	157.0	18
China	62.2	52.3	114.5	13
Australia	37.1	39.3	76.4	9
India	56.1	4.5	60.6	7
European Union	5.1	51.0	56.1	7
Africa	31.5	0.2	31.7	4
Other	55.2	72.2	127.4	15
World Total	404.8	456.2	860.9	100

Source: BP Statistical Review of World Energy 2012, http://www.bp.com/liveassets /bp_internet/globalbp/ globalbp_uk_english/reports_and_publications/statistical_ energy_review_2011/STAGING/local_assets/pdf/ coal_section_2012.pdf.
Notes: Percentages may not add to 100% because of rounding.

Global Coal Use on the Rise

As a result of increasing electricity demand, global coal consumption has steadily risen over the last decade, an era of economic recession, which portends opportunities for increases in U.S. coal exports. The IEA projects that by 2017 coal use will be on par with oil as the top fuel in the world's energy mix (see Figure 2).[8]

Worldwide, coal is the largest contributor to primary energy growth, accounting for 65% of the increase in primary energy between 2010 and 2011.[9] And, China's coal consumption was the largest contributor to the global growth in coal, up almost 165 Mtoe, in 2011 for about a 10% increase.[10] Most of China's coal consumption is for its rapidly expanding electricity generation sector (see Figure 3).

The global coal market in 2011 was over 3,700 Mtoe or 8.2 billion short tons (BST), up about 68% since 2000.[11] This represents an annual average growth rate of roughly 5%, the highest among fossil fuels. China dominates the global coal market, accounting for almost half of global coal consumption in 2011. For comparison, the world's second- and third-largest coal consumers, the United States and India, account for 13% and 8% of global consumption, respectively. The European Union consumed about 8%.

China's growing consumption of coal outpaced its production, taking that country from being a net coal exporter to a net importer. In 2000, China was the world's second-largest coal exporter, behind Australia. China started importing more than it was exporting in 2009 and became the world's second-largest importer, after Japan, according to EIA statistics.[12] In 2010, net imports to China represented 5% of China's domestic consumption.

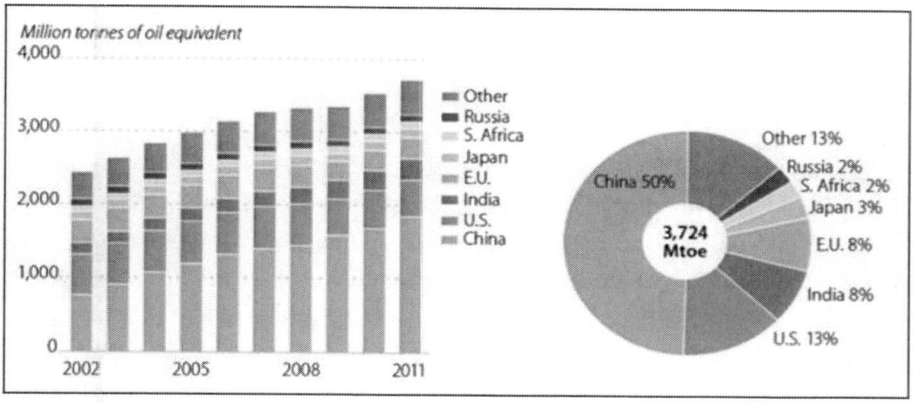

Source: BP Statistical Review of World Energy, 2012.

Notes: Pie chart based on 2011 data.

Figure 2. Global Coal Consumption, 2002-2011.

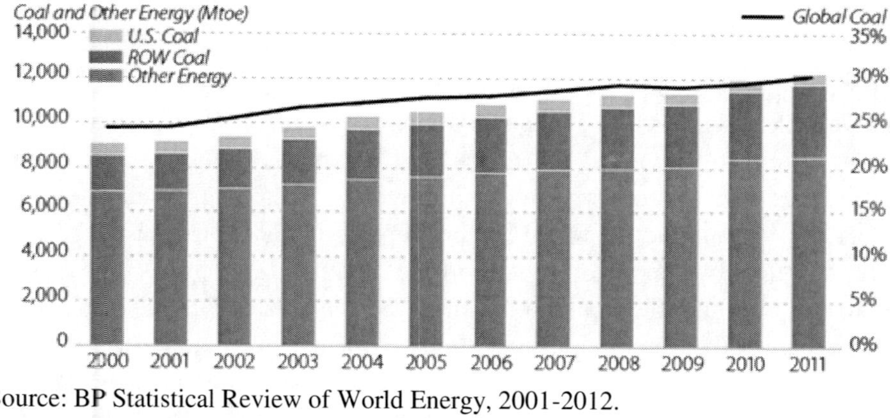

Source: BP Statistical Review of World Energy, 2001-2012.

Notes: Primary energy comprises commercially traded fuels. Units = million tonnes of oil equivalent (Mtoe).

Figure 3. Global Coal Growth as Part of Primary Energy Consumption, 2000-2011.

International trade is more limited in coal than other fossil fuels. Roughly 15% of coal—1.2 billion short tons (BST) is traded globally versus 60% of oil and 30% of natural gas.[13] The largest coal consumers are also the world's largest producers—China, the United States, and India. However, outside of China and India, most of the growth in global coal production has come from major coal exporters Indonesia (likely to curtail exports in the coming years for domestic use), Australia, Russia, and Colombia. Essentially, they are filling the space left in the market as China swung from being a net exporter to a net importer.

Globally Traded Coal

Global trade in coal has averaged about 15% of world production, with the United States ranking among the top five exporting countries in 2010 through 2012.[14] Australia, Indonesia, Russia, South Africa, and Colombia have also ranked in the top five coal exporting countries over the last five years. Meanwhile, Japan, China, South Korea, Taiwan, India, and Germany have ranked in the top five importing countries during those five years. Most notably, China climbed to be the second-largest importer in 2009, surpassing Japan for the top spot in 2012.[15]

Global Prices and Contracts

The value of coal varies according to certain characteristics—energy content, ash level, sulfur content, moisture, and volatiles—while contracts include factors such as use, coal quality, length of contract, delivery terms, and payment options.[16] These variables have made comparison of coals for export more difficult than for other hydrocarbons, which tend to be more uniform.

Coal prices around the world reached a peak in 2008 and then fell sharply with the global economy, as did oil and natural gas prices. As prices rebounded in 2009, the gap between the U.S. benchmark Central Appalachia coal price and other international coal prices widened (see Figure 4).

As more U.S. coal enters the global market, the additional demand will likely raise domestic prices while contributing low-cost coal to the world market. All else being equal, the additional U.S. low-cost coal might result in more coal being consumed worldwide.

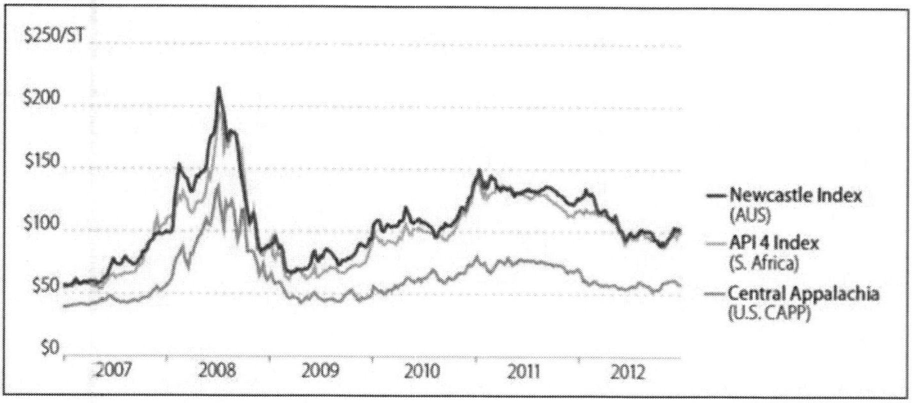

Source: Morgan Stanley.

Notes: All prices are FOB (free on board) at respective terminals. The API 4 Index price is the benchmark price reference for coal exported from South Africa's Richards Bay terminal. Units = U.S. dollars per short ton ($/ST). FOB prices are for the product plus the cost of loading the coal on a vehicle or vessel.

Figure 4. Global Coal Prices; 2007-2012.

U.S. EXPORTS: TRENDS POINT UP

The United States has been exporting coal since the late 1800s.[17] From 2003 to 2012, U.S. coal exports rose over 200%, mainly driven by competitive production costs, global demand, and lower prices, among other factors[18] (see Figure 5). Coal exports comprised 12% of U.S. coal production in 2012.

The Recent Past of U.S. Coal Exports

The recent U.S. economic crisis caused a sharp decline in U.S. 2009 exports over 2008, but exports are again trending up, more than doubling since 2009. U.S. coal exports were aided by a drop in value of the U.S. dollar against other currencies, including those of other major coal exporting countries, such as Australia, Indonesia, and Russia. Exports of both metallurgical coal and thermal coal from the United States have been increasing since 2009, rising almost 87% and 73%, respectively.[19] Prior to 2009, exports of both had been on the rise, but the economic crisis stifled that growth.

In 2011, U.S. coal exports broke 100 MST for the first time since 1992 and in 2012 reached 126 MST, surpassing their previous peak of almost 113 MST in 1981. The value of U.S. coal exports has increased, rising from under $10 billion in 2010 to almost $16 billion in 2011, according to EIA data (see Figure 6).[20]

Exports by State

Figure 7 shows a breakdown of coal exports by states. West Virginian coal accounted for the largest portion of U.S. coal exports. Over the last five years, coal from the Appalachian region fell as a percent of the total amount of coal exported, but has grown almost 70%, in absolute terms, since 2007.[21] Other key Appalachian region coal exporters include Pennsylvania and Virginia. Western states, led by Montana, comprised the second-largest share of exports, 21%, in 2011. The Interior region is up the greatest amount in percentage terms, rising from 1.5 MST to 7.7 MST during the same time period.

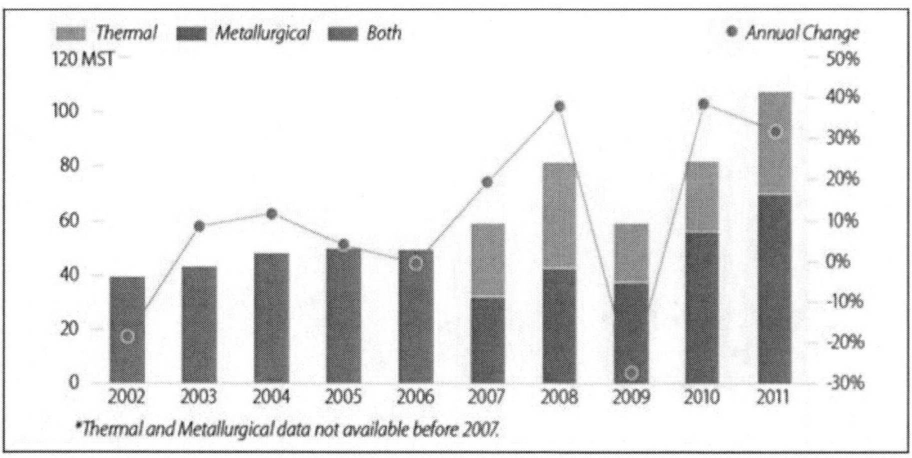

Source: U.S. Energy Information Administration, http://www.eia.gov/totalenergy/data /annual/showtext.cfm?t= ptb0705.

Notes: Units = million short tons (MST).

Figure 5. U.S. Coal Exports; 2002-2011.

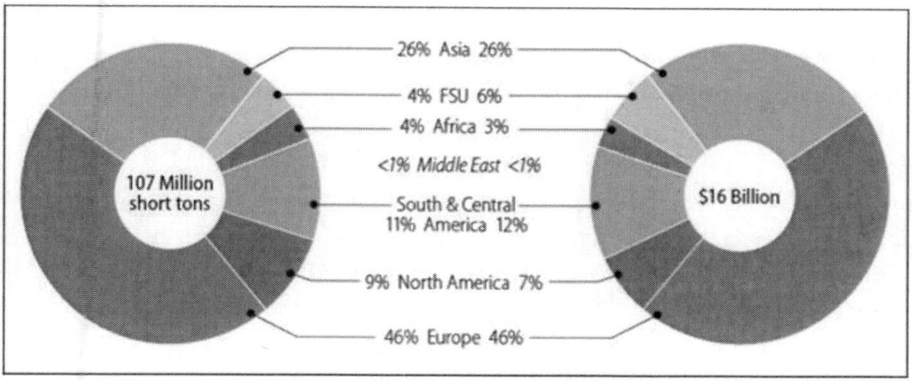

Source: U.S. Energy Information Administration, Historical U.S. Coal Exports by Year and Quarter (2002 – 2011), http://www.eia.gov/coal/data.cfm#imports.

Notes: FSU = former Soviet Union.

Figure 6. U.S. Coal Exports and Trade, 2011.

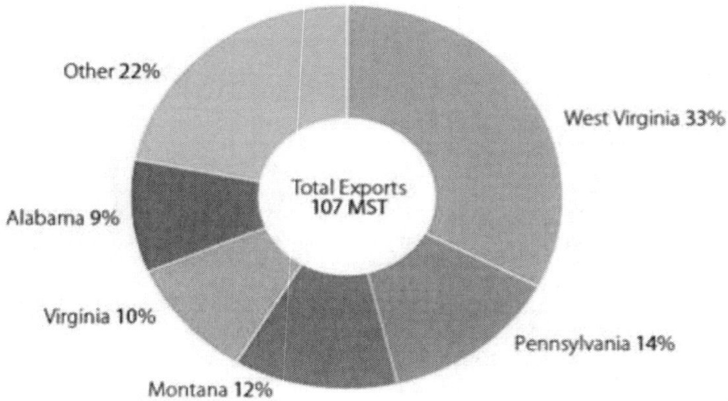

Source: U.S. Energy Information Administration, http://www.eia.gov/coal/distribution /annual/pdf/ o_11foreign.pdf.

Notes: Units = million short tons (MST).

Figure 7. U.S. Coal Exports by State, 2011.

Domestically, other factors have contributed to the rise in U.S. coal exports since 2009. U.S. natural gas prices have trended significantly lower since peaking in mid-2008 because of the economic crisis and the introduction of large amounts of shale gas to the market. Coal and natural gas compete directly in electric power generation. Historically, coal plants were more expensive to build, but the fuel cost was cheaper. Natural gas prices in 2012 have dipped below coal prices on an energy equivalent basis, altering the industry's perception. Coal and natural gas reached parity in electricity generation in April 2012 for the first time. But are cheap natural gas prices a sustainable trend? Electricity generation accounts for 93% of coal consumption, and the decrease in coal generation contributed to lower coal prices relative to international markets. The trend for gas prices is projected to continue to decrease the competitiveness of coal as a fuel for electric power generation in the U.S. market, according to the EIA. Therefore, in order to maintain production levels, coal companies are looking more to foreign markets and exports as a growing segment of their market.

Table 3. U.S. Exports to Selected Countries, 2011 in MST

	Metallurgical coal	Steam Coal
Western Hemisphere	**14.2**	**7.2**
Canada	4.2	2.7
Brazil	8.3	0.3
Other	1.7	4.2
Europe	**34.3**	**19.6**
France	1.8	2.3
Germany	2.1	2.6
Italy	4.5	1.1
Netherlands	6.5	4.3
Ukraine	4.7	0.1
UK	2.7	4.2
Other	12.0	5.0
Asia Pacific	**19.7**	**7.8**
China	4.6	1.0
India	3.8	0.7
Japan	6.2	0.7
South Korea	5.1	5.4
Other	0.0	0.0
Africa	**1.4**	**2.7**

Source: U.S. Energy Information Administration, *Quarterly Coal Report*, April 18, 2012, http://www.eia.gov/coal/ production/quarterly/.

Of the 107 MST exported in 2011, about 54 MST went to Europe and 28 MST went to Asia. Asia's largest recipients include South Korea, Japan, and China (10 MST, 7 MST, and 6 MST respectively).[22] Nearly 70 MST of U.S.

exports in 2011 were metallurgical coal (coking coal), up from 56 MST in 2010. Steam (thermal) coal accounted for 38 MST in 2011, up from 26 MST in 2010. Metallurgical coal has a higher energy content and is used to make coke used in making steel. It is mostly produced in Appalachia, whereas most steam coal used for heating and power generation is produced from the Illinois Basin and the Powder River Basin (PRB).

Most U.S. coal is exported through Mid-Atlantic ports in the Eastern U.S. customs district (see Table 4), particularly Norfolk, VA, and Baltimore, MD. Nearly 40% of U.S. exports passed through Norfolk, and 34 MST of 41 MST exported were metallurgical coal. This coal was primarily from West Virginia, Pennsylvania, and Virginia. A large amount of coal is also exported out of the Southern ports of Mobile, AL, and New Orleans, LA. Western exports of coal were 7.4 MST, of which nearly 5 MST passed through the Seattle, WA, customs district.[23]

Table 4. U.S. Coal Exports by Customs Districts in MST

	2010	2011
Eastern District	**47.5**	**62.1**
Baltimore, MD	13.9	19.3
Norfolk, VA	32.0	41.0
Other	1.6	1.9
Southern District	**19.8**	**32.6**
Mobile, AL	9.7	10.1
New Orleans, LA	9.4	21.4
Other	0.7	1.1
Western District	**5.1**	**7.4**
Seattle	3.5	4.9
Other	1.6	2.5
Northern District	**9.2**	**4.9**
Total Exports	**81.7**	**107.3**

Source: U.S. Energy Information Administration, *Quarterly Coal Report*, April 18, 2012, http://www.eia.gov/coal/ production/quarterly/.

Notes: May not sum to total due to rounding. Customs districts incorporate a wider geographic area than just the place name. For instance, the Seattle customs district includes Aberdeen, Spokane, and Blaine, WA.

With greater energy content per pound, metallurgical coal can be more attractive to ship over longer distances. Further, global metallurgical coal demand has been boosted in recent years due to the rapid economic growth in emerging markets, which involved infrastructure and industrial expansion, creating demand for steel. According to the IEA, global metallurgical coal consumption grew by 87% between 2000 and 2010, versus 70% growth of steam coal and 12% growth in brown coal or lignite,[24] with China accounting for 80% of global metallurgical coal consumption growth between 2000 and 2010.[25] For context, according to the IEA, metallurgical coal makes up 14% of global coal consumption but more than 30% of the international coal trade.[26]

The Future of U.S. Coal Exports

There are several key factors likely to influence how much coal will be exported from the United States. Several export terminal projects have been proposed by 2016, such that the United States could be exporting more than double its current coal exports. Projects in the Pacific Northwest have attracted much of the attention, even though the Northeast or Mid- and South Atlantic ports continue to account for most exports. Additionally, the Gulf Coast export facilities may be in a good position to expand capacity and incrementally increase exports. In its 2013 Annual Energy Outlook Early Release, EIA's base case projects that total U.S. coal exports will rise steadily starting in 2013 through 2038, growing over 50% during the time period. Also, it is worth noting that EIA's coal export base case has been revised significantly higher since its 2012 forecast.[27]

Powder River Basin: A New Focus for Exports

There is growing interest in exporting Powder River Basin (PRB) coal. Low production costs, large reserves and production, and low sulfur content can make PRB coal attractive to domestic and foreign electric utilities. However, PRB coal has greater moisture content and lower energy content per pound, making it more expensive to transport on a per unit of energy basis.

Today, nearly all coal produced from the PRB is consumed within the United States (mostly in the Midwest), and it is used almost entirely for electric power generation. PRB coal has a relatively low sulfur content, which makes it easier for coal-burning utilities to meet Clean Air Act (P.L. 101-549) restrictions on sulfur dioxide emissions. The PRB sits largely on federal lands in Montana and Wyoming, and according to estimates accounts for more than 90% of the total coal produced from the two states. In 2010, the latest year for export data by state, Wyoming and Montana produced just over 487 MST of coal, of which less than 1% was exported.[28]

While the vast majority of PRB coal stays within the United States, exports of PRB coal have increased in 2011 as total U.S. exports have increased significantly. Coal distribution data from EIA for 2011 show that 95% of PRB coal has been distributed around the United States, mostly for electric generation.[29] Note that while PRB's low sulfur content may help utilities meet EPA sulfur regulations, new EPA regulations, such as those on hazardous air pollutants, may still reduce domestic PRB demand if overall coal-fired electricity declines.[30]

PRB coal is exported primarily from Canadian terminals at Roberts Bank (near Vancouver, British Columbia) and Ridley Terminal at Prince Rupert, British Columbia. PRB coal is transported to both facilities for export via railway. However, the Canadian export terminals have reached capacity.[31]

Although the Canadian export facilities have plans for expansion that may better accommodate U.S. exports,[32] PRB coal producers have been searching for a potential domestic export link to the growing Asian market. That link appears to be through the Pacific Northwest. Three port terminal projects for exporting coal in Washington and Oregon have permit applications pending before the U.S. Army Corps of Engineers (the Corps):[33]

- Gateway Pacific Terminal (GPT) and Rail Expansion, Cherry Point, WA;[34]
- Millennium Bulk Terminal (MBT), Longview, WA;
- Coyote Island Terminal (CIT) at the Port of Morrow, Boardman, OR.[35]

Several other projects have been discussed, but it is unclear at this point how many or when any additional projects may move forward.

Gulf Coast Export Capacity Likely to Grow

Although increasing coal exports from the Pacific Northwest are attracting the most attention, the Gulf Coast may be a more likely candidate to add export capacity in the short term. The region has more existing capacity to build from and less local opposition to increased exports.[36] Over the next five years, over 40 MST of proposed expansions and new projects will be evaluated, particularly for Illinois Basin coal (IB) and PRB coal. Transporting IB coal by barge to export terminals in the Gulf, especially New Orleans, is cheaper than moving other coal to the coasts by rail. Located in the central United States, the Illinois Basin ranges across Illinois, Indiana, and western Kentucky. The coal produced in this region is bituminous coal with relatively high sulfur content. Production within the state of Illinois (23.9 MST) has been a key driver of the basin production increase over the first half of 2012.

Exports of IB coal have been rising but only account for about 13% of IB production, which has also been increasing. IB coal is cost competitive on the international market, with most exports going to the United Kingdom and the Netherlands (approximately 3.8 MST and 1.4 MST, respectively), but usually requires a discount because of its high sulfur content. China was the third-largest recipient, with 1.2 MST. Coal exports to international markets from Illinois doubled from 2010 to 2011 (2.5 million tons in 2010 to 5 million tons in 2011). In addition, domestic demand for coal from the Illinois Basin, particularly in Illinois itself, increased as a result of a shift in demand toward the Illinois Basin's low-cost, high-sulfur coal and away from Central Appalachia's high-cost, low-sulfur coal. Domestic utilities that have added scrubbers can burn high-sulfur coal while remaining in compliance with existing requirements to reduce sulfur dioxide emissions. Because of the relatively low cost of Illinois Basin coal and its use in larger, efficient plants with modern pollution control equipment, its producers were less affected by recent low natural gas prices.

East Coast Exports

Coal shipments from East Coast terminals of Appalachian coal have dominated U.S. coal exports, accounting for almost 60% and 71%, respectively, in 2010 and 2011. Appalachian coal, led by production in West Virginia, has declined as a percent of overall U.S. coal exports over the last five years, but has grown almost 70% in absolute terms. Coal exports from Pennsylvania and Virginia are also up, 169% and 65%, respectively, in the same period.

Multiple projects are being reviewed to increase East Coast export capacity of Appalachian coal. Although there are proposals for constructing new facilities, most companies are looking to expand their existing terminals: expansions tend to have fewer regulatory requirements and lower cost.

Exports: Transportation Issues

Getting Coal to Ports

Given coal's low value-to-weight ratio, the large volume and non-time sensitive nature of shipments, and the long distances typically required for movement from coal mine to port, rail and barge offer the most economical means of transporting coal for export. In the West, a coal train of 120 cars can carry over 14,000 tons of coal (118 tons per car). Coal trains in the East are somewhat shorter in length. One barge holds 1,570 tons of coal but 15 barges (holding 23,550 tons) are often tied together in a tow through locked portions of inland rivers. On the Mississippi River below St. Louis, where there are no locks 30 to 40 barges can make up a single tow of coal destined for terminals on the Mississippi River, where it is loaded aboard oceangoing ships for export from Louisiana.

Rail transport from mid-Atlantic mines to Norfolk, VA, has historically been the leading pathway for U.S. exports of coal. Secondary pathways include movement of Alabama coal to Mobile by rail and/or truck, barge transport of Midwest and Ohio Valley originated coal to Mississippi River Delta ports, and rail transport of Pennsylvania coal to Baltimore (see Appendix A, Figure A-1 for U.S. coal regions, and infrastructure). Some U.S. coal is exported from Ohio across the Great Lakes to Canada. Western U.S. coal is also shipped by rail to Duluth, MN, and then through the Great Lakes to Quebec, where it is moved to large oceangoing ships for export to Europe.[37]

An obvious way to export PRB coal to China, Japan, or Korea would be through ports in Washington, Oregon, or British Columbia, Canada. Most coal from the PRB is currently moving east or south by railroad. A small fraction moves by rail to Washington and Oregon for domestic use and to Vancouver and Prince Rupert, BC, for export. While not currently used heavily for coal transport, east-west rail lines are the main pathways for carrying Pacific Northwest port traffic. However, even though coal trains are slower and less time-sensitive than trains carrying other commodities, they are cumbersome to move to sidings due to their exceptional weight, and can interfere with other traffic.[38]

Coal terminals at the California ports of Richmond, Stockton, Los Angeles, and Long Beach can be used to export coal from other western mines to Asia. Although the routing is more circuitous, PRB coal could be exported to China, Japan, or Korea by rail through Texas ports or through Mississippi River Delta ports (possibly in combination with barge transport from upriver terminals). These routes avoid the Rocky Mountains and more severe winter weather, and the tracks are in place to handle heavy coal traffic. The central location of Gulf ports also allows for some hedging against coal sourcing (western, central, or Appalachian mines) as well as Trans-Pacific versus Trans-Atlantic destination markets (India is closer to Gulf ports via a Trans-Atlantic crossing and passage through the Suez Canal). Charleston, SC, and

Philadelphia, PA, on the East Coast could become additional conduits for export of Appalachian coal.[39]

Most rail freight travels under rates negotiated between shippers and railroads. Typically, rates are specified in confidential contracts that may last for months or years, and may include other provisions such as discounts to the shipper for exceeding a certain volume during a given period or penalties to the railroad for late delivery. A coal contract might cover shipments from a single mine or to a single power plant, but it could also cover multiple origination or destination points.

Under existing law, governments at all levels have limited influence over the type or quantity of cargo railroads carry on their networks. Railroads' construction plans are generally not subject to state or local permitting laws,[40] and approval by the federal Surface Transportation Board is necessary only for construction of new railroad rights-of-way and railroad mergers or acquisitions and generally not for improvements to existing lines.[41] A railroad's share of the cost when improving highway-rail grade crossings is limited by federal law.[42]

PRB Rail Tracks

As the nation's largest source of coal, the PRB has been the focus of much attention. Union Pacific and BNSF Railroads share a right-of-way out of the PRB—the "joint line."[43] It is 103 miles in length and is mostly triple tracked with a fourth track on segments with hills. It is the busiest rail track in the world,[44] carrying most of the PRB coal from the mines to points south and east. In May 2005, two derailments caused by ballast[45] deterioration from heavy rains and coal dust (the railroads contend coal dust prevents proper drainage) severely interrupted PRB shipments. The railroads made extensive repairs and upgrades to the line but this took two years to complete. To keep coal dust from blowing off cars in transit, railroads are mounding or shaping the coal more uniformly when loading (to reduce wind resistance), and applying a surfactant as a sealant. The railroads and coal shippers have disputed who should pay for these measures.[46] Plans for an additional line into the PRB from the north, the Tongue River Railroad, are under review by the Surface Transportation Board (STB).[47]

Barge Infrastructure Issues

A growing concern of barge operators has been the reliability of locks because of more frequent scheduled and unscheduled maintenance by the Army Corps of Engineers. Many of the locks were originally constructed in the 1930s or 1950s. The cost of keeping locks functioning has increased significantly in recent years, and has surpassed current financing mechanisms. This includes a 20-cent-per-gallon federal tax on barge fuel that offsets about 10% of the federal cost of providing inland waterway infrastructure. The insufficiency of funds generated by the fuel tax would be exacerbated if demand for coal were to decline substantially. In a period of constrained federal budgets, Congress is evaluating alternatives for financing inland waterway infrastructure.[48]

Harbor Depths

Most coal shipped overseas from the United States is carried in Panamax vessels. These vessels are of a size limited by the dimensions of Panama Canal locks. They can carry about 60,000 tons of coal to keep their draft from exceeding 40 feet, the allowable draft depth of Panama Canal locks. Some U.S. export coal is loaded into Capesize vessels (which are too large for the Panama and Suez Canals and therefore must sail around the capes of South Africa and South America). Capesize vessels have a capacity of 90,000 to 200,000 short tons and a draft in the range of 48 to 56 feet. The Capesize vessels loading coal at Norfolk and Baltimore are in the smaller range of this class with drafts up to 50 feet. The expansion of the Panama Canal in early 2015 will allow smaller Capesize vessels with 50-foot drafts to pass through, but many U.S. ports and foreign unloading ports have insufficient depth to handle this type of ship. As discussed in greater detail in the U.S. Export section of this report, below, the three proposed port terminals with permit applications would have capacities of 54 million metric tons (GPT), 44 million metric tons (MBT), both in Washington State, and 8 million tons (CIT) in Oregon.

The economies of scale that can be achieved with a ship carrying twice as much cargo (Capesize versus Panamax) can significantly reduce the per ton cost of ocean transport. Only Puget Sound and Los Angeles area terminals have the depths to handle large Capesize vessels. Table 5 below lists the depths of ports that currently export coal or have been mentioned as possibly handling coal in the future. With the exception of Puget Sound ports (Bellingham), Southern California ports, and Canadian ports, which are all naturally deep harbors, all of the ports listed require regular maintenance dredging (ranging from continuous to yearly to episodic) in order to maintain channel depths and widths. Congress is debating the appropriate level of funding for harbor maintenance.[49]

Table 5. Harbor Depths of Current or Proposed Coal Ports by Region (depth in feet at mean low water, if tidal)

Harbor	Depth (in feet)
Canada	
Prince Rupert, Canada	75
Vancouver, Canada	68-75
Quebec, Canada	50
Western	
Bellingham, WA (Cherry Pt.)	96
Lower Columbia River, OR and WA	43
Coos Bay, OR	37
Richmond, CA	38
Stockton, CA	35
Los Angeles, CA	72
Long Beach, CA	50
Gulf Coast	
Corpus Christi, T)(	45
Houston/Galveston, T)(	45
Lower Mississippi River, LA	45
Mobile, AL	45
Eastern	
Charleston, SC	45
Norfolk/Hampton Roads, VA	50
Baltimore, MD	50
Philadelphia, PA	40
Other	
Great Lakes	28-32

Source: U.S. Army Corps of Engineers, port and terminal websites.

Notes: Coal ships generally can wait for high tide before transiting a harbor. Ships require two to three feet of under-keel clearance.

Port Development and Maintenance

Generally, commercial port development is initiated and governed at the local level by harbor communities via a port authority, not by the federal government. A port authority could be a function of state, county, or city government. Federal jurisdiction over port development is primarily concerned with waterside infrastructure and operations. The U.S. Army Corps of Engineers is responsible for dredging shipping fairways and maintaining breakwaters and jetties.

Reflecting local concerns, Congress authorizes the depths of harbors and determines the Corps' annual budget for harbor maintenance and operations.[50] Deepening a port is funded from the General Treasury, while Corps maintenance costs are paid with an *ad valorem* tax on imported and domestic port cargo.[51] Non-federal stakeholders are responsible for dredging ship berths and other non-federal channels in ports. The development of landside terminal infrastructure, such as coal unloading/loading facilities, is typically financed by private industry but may be assisted by the port authority. Terminal property could be privately owned by a shipper or owned by the port authority and leased to a terminal operator.

Shipping Rates

Shipping rates in the dry bulk sector are notoriously volatile, depending on global supply and demand conditions. A new vessel requires two or more years to build and this lag can lead to oversupply or shortage conditions. Dry bulk vessels are the simplest big ships to build and are favored by new shipyards in boom years. Such ships are versatile, geographically and across commodities. Coal competes with iron ore for Capesize vessel capacity and with cement, steel, fertilizer, woodchips, salt, and grain for Panamax vessel capacity. Ocean freight rates, especially for longer trans-Pacific voyages, are significant relative to the value of coal, and thus fluctuations in dry bulk vessel rates will affect the price competitiveness of U.S. coal compared to foreign sources. Currently, dry bulk rates are low but their range of decline in recent years also indicates their upswing potential. From 2010 to 2011, dry bulk rates generally decreased by 70% to 80%.[52]

COAL EXPORTS AND THE ENVIRONMENT

Support and Opposition to Increased PRB Exports

Representatives from state and local agencies along the potential coal transport corridor, particularly in Washington and Oregon, as well as industry, community, and environmental groups, have raised certain concerns over port terminal projects that would allow for increased export of PRB coal through the Pacific Northwest. Opponents, including several environmental groups, have argued that increased train and barge traffic would have significant adverse impacts to the human, natural, and cultural environment.[53] Environmental groups have recently filed a lawsuit against the BNSF Railway

Company and several coal companies under the Clean Water Act in the U.S. District Court for the Western District of Washington.[54] Project supporters have argued that the projects would create or maintain jobs in the construction, mining, and transportation industries and bring increased tax revenue to the states. Supporters such as the coal companies and labor unions also argue that many potential adverse impacts can be mitigated.

Some stakeholders are taking a wait-and-see approach—recognizing potential benefits, but waiting to see what the cumulative effects of the various port terminal projects may be, including mitigation measures that would be implemented, before deciding whether potential benefits outweigh adverse impacts.

Stakeholder concerns over the cumulative impacts to communities, tribes, states, and regions primarily pertain to effects of increased rail and barge transport of PRB coal through the Pacific Northwest. Considering the pending and potential projects, it would appear that annual PRB coal export capacity could expand to exceed 100 million short tons. Rail industry representatives have stated that such an increase in export capacity could result in 8 to 16 additional trains per day. The coal would arrive through the Columbia River Gorge by rail from the PRB in Montana and Wyoming, and continue by rail and/or barge to export facilities in Washington and Oregon.

The potential for an increase in mile-and-a-half long coal trains has generated concern regarding potential adverse impacts from increased coal dust emissions to air and deposition to land or surface water; traffic congestion and increased wait times at rail crossings; potential derailments or delays to emergency response vehicles; and noise and vibrations high enough to cause structural damage. Dredging and construction activities in waterways to accommodate increased barge traffic and deep-water vessels have also generated concern over potential impacts to fisheries, marine ecosystems, or endangered species habitat.[55] Also, some stakeholders have cited broader global climate change impacts associated with increased coal burning in Asia.

Project proponents argue that many of the potential adverse effects, particularly those related to increased barge and rail transport, can be mitigated or largely eliminated.[56] However, some stakeholders have expressed concern that the potentially broader, cumulative impacts of pending and possible future projects will not be adequately identified and, thus, not be addressed.

Efforts to explicitly require or negotiate agreements with coal producers and transporters to mitigate effects of transporting PRB coal through the Pacific Northwest will likely occur within the context of existing local, state, tribal, or federal requirements applicable to individual port terminal projects.

Since permit applications have been submitted to the Corps, state governors, local, state, tribal, and federal agencies, community and environmental organizations, and individual members of the public have submitted comments to the Corps urging them to ensure that indirect and cumulative impacts of the projects are fully identified and considered before they make a final decision whether to issue those permits. The framework within which the Corps will identify and analyze indirect and cumulative project impacts will comply with the National Environmental Policy Act (NEPA).

Opponents of the coal export projects have asked the Army Corps of Engineers to conduct a more comprehensive region-wide environmental impact statement that would look at impacts ranging from mining to burning U.S. coal overseas. This viewpoint was rejected by the Corps of Engineers, stating that an analysis of the potential broad-based impacts, i.e., from mining through burning U.S. coal overseas, was beyond their scope and jurisdiction. However, the Washington Department of Ecology has plans for a comprehensive review of environmental, transportation, health, and climate issues before the coal ports can be built. Scoping the environmental review for the proposed terminal near Longview, WA, being coordinated by the Department of Ecology, the Army Corps of Engineers, and Cowiltz County, began on August 16, 2013. The three agencies plan to issue a joint environmental impact statement (EIS).

Coal Exports and Greenhouse Gas Emissions

Two perspectives may shape policy makers' consideration of greenhouse gas emissions (GHG) emissions and the coal export issue: a legal view and a physical/environmental view. From the legal perspective, some might ask whether coal exports could put the United States at risk of violating obligations under existing or future domestic laws or international agreements to address climate change. From a physical view, regardless of legal requirements, a question is whether increasing coal exports may exacerbate climate change and the risks it poses by increasing U.S. and/or global emissions of greenhouse gases.

GHG Emissions from Coal Production and Combustion

Coal production and use result in emissions of greenhouse gases (GHG) comprised mostly of carbon dioxide (CO_2) and methane (CH_4). These gases are believed to contribute to rising global average temperatures and other climate changes.[57] Methane is primarily associated with fossil fuel production while carbon dioxide results primarily from coal combustion.

Coal seams contain methane, which typically is released when coal beds are disturbed. The methane emissions continue for many years after mining ends. Globally, coal production is estimated to have contributed about 10% of the higher concentrations of methane in the atmosphere since the Industrial Revolution began. Methane is roughly 56 or 21 times more potent (ton-for-ton) than carbon dioxide over a 20-year or 100-year period, respectively. (The index depends on the time period of policy-making interest: CO_2 remains in the atmosphere for hundreds of years once emitted, while methane in the atmosphere oxidizes to CO_2 or other compounds with a half-life of about a decade.) Methane emissions may be effectively abated, frequently for safety and their energy value.

Coal combustion releases carbon dioxide, the most pervasive GHG. Globally, coal contributes an estimated 43% to energy-related emissions of CO_2, which in turn is about three-quarters of all GHG emitted by human activities. Abating carbon dioxide emissions requires capture and sequestration technologies not yet commercially available.

Legal Implications: No Binding Obligations, but Future Ones Could Be Exacerbated

Rising coal exports would not compromise the ability of the United States to meet any existing, enforceable climate-related obligations of the U.S. federal government. Domestically, the Environmental Protection Agency (EPA) has set certain regulations to control methane emissions, but not on coal mines. The EPA has proposed performance standards for carbon dioxide emissions from large combustion units,[58] but these would not apply to exported coal.

In the international context, current federal commitments to reduce coal-related GHG emissions are not legally enforceable. For example, under the United Nations Framework Convention on Climate Change (UNFCCC), the

United States adheres to an objective of avoiding dangerous human interference with the climate system, and to develop national plans that would reduce national GHG emissions. The United States has not agreed to legally binding, quantitative emissions caps. President Obama in 2009 stated a policy that the United States should reduce its GHG emissions to 17% below 2005 levels by 2020. However, there are no binding mechanisms for the international community to enforce this U.S. policy.[59]

In the future, rising coal exports could make it more challenging to reach any potential, absolute GHG target, whether set domestically or internationally. For the United States, on the one hand, increasing production of U.S. coal would tend to increase related methane emissions above what would otherwise occur. On the other hand, coal exports would tend to raise domestic coal prices and thereby tend to curb U.S. coal use and related emissions. The balance between increases and decreases is unknown.

U.S. coal exports could make it more difficult for other countries to meet absolute GHG targets. Reportedly, European utilities have increased imports of relatively inexpensive coal from the United States, raising their GHG emissions and complicating their efforts to meet the European Union's legally binding GHG targets under European Union law and the Kyoto Protocol of the UNFCCC.

U.S. coal exports to China, India, and other rapidly growing coal consumers would likely affect efforts to persuade them to commit to reducing their GHG emissions. Past Congresses have resolved that binding commitments from such countries must be made for the United States to bind itself to GHG emissions under the UNFCCC.[60] In this sense, increasing U.S. coal exports would likely impair U.S. persuasiveness and cooperative efforts to negotiate a legal international framework that includes GHG reductions from China, India, and other countries.[61]

Physical Implications: Rising GHG Emissions Contribute to Climate Change

Increasing U.S. coal exports would likely contribute to rising global emissions of GHG, by increasing availability of coal and stimulating greater coal consumption due to lower market prices. Higher GHG emissions would add to suspected human-induced climate change globally as well as to local and transboundary air pollution and related health risks.

As noted above, U.S. coal production for export would increase U.S. emissions of production-related methane above what would otherwise occur, but could increase or decrease U.S. combustion-related emissions of carbon dioxide.

Current coal exports come primarily from the Appalachian region, which generally has higher potential emissions of methane than western surface mines.

Though producers of Appalachian coal plan to expand exports,[62] much of the anticipated increase in exports would be produced in the Powder River Basin.[63]

The net effect is likely that increasing U.S. exports of coal could contribute to rising emissions of greenhouse gases (carbon dioxide and methane) as well as other pollution, although the increase could be small. No published study has thoroughly quantified this complicated question.

Table 6. Conceptual Legal and Climatological Implications of Increasing U.S. Coal Exports

	U.S. Domestic	Global
Legal Implications	None currently at federal level though possible in some states; could make setting any future, fixed GHG target more difficult.	Would make meeting existing, non-binding objectives more difficult.
Climatological Implications	Would increase coal production, processing, and transportation emissions in the United States. More international competition could raise or lower U.S. coal prices, tending to lower or raise (respectively) U.S. GHG emissions and consequent impacts on climate.	Would increase global emissions if U.S. exports increase available supplies and/or help suppress coal prices internationally; lower coal prices could also suppress investments in efficiency, nuclear, or other energy alternatives.

Source: CRS.

A Role for NEPA[64]

Broadly, NEPA requires federal agencies to consider the environmental impacts of their actions before a final decision is made regarding that action.

Consideration of a proposed project's environmental impacts is described in the appropriate NEPA document. For projects anticipated to have a "significant" impact on the quality of the human environment, an environmental impact statement (EIS) must be prepared. Projects with uncertain impacts require preparation of an environmental assessment (EA) to determine whether the impacts would be significant.[65] The Corps has identified permit issuance among those actions normally requiring an EA, but not necessarily an EIS.[66]

However, of the three Pacific Northwest coal terminal projects, the Corps has determined that an EA will be prepared for the Coyote Island transfer terminal and that the GPT and MBT will have significant impacts, requiring the preparation of an EIS.

During the project "scoping" phase of the NEPA compliance process, the Corps must clarify significant issues that will be analyzed in depth in the EIS. The scoping phase is currently underway for the GPT and MBT projects. As a result, the full range of issues that will be analyzed in each EIS is not yet known.

However, in determining the scope of an EIS, the Corps is required to consider connected, cumulative, or similar project-related actions and the resulting direct, indirect, and cumulative effects of those actions.[67]

Permit applications pending before the Corps for export terminal expansion projects in Oregon and Washington may be considered connected or similar actions that would result in certain indirect or cumulative impacts. What is uncertain, however, is how broadly the Corps may define those terms. For example, it is unclear whether the Corps may identify increased train traffic out of Montana or Wyoming as an indirect effect of increasing coal export capacity related to permit issuance for the GPT or MBT projects. Also, given the current limits to commercial rail capacity in Washington and Oregon, it is not known whether the Corps would identify the potential need to expand rail capacity as a reasonably foreseeable indirect effect of port terminal permit approval.

The potential need for new rail construction or the expansion of existing rail facilities may be particularly relevant if such activities were subject to approval from the U.S. Surface Transportation Board (i.e., a federal action subject to NEPA).

While it is difficult to determine the full range of project-related actions and resulting impacts that the Corps may include in the EA and EISs currently being prepared, a letter from the Corps indicates its approach to the environmental review process for the proposed projects. In response to a letter from

Oregon Governor John Kitzhaber regarding concerns over the cumulative impacts of port terminal projects proposed in Washington and Oregon, the Corps states, in part:

> The effects that the Corps will consider in the review of each project proposal will include the specific activity requiring a DA [Department of the Army] permit, and those portions of the entire project over which there is sufficient federal control and responsibility to warrant Corps NEPA review. Based on existing law and DA regulations, the Corps will extend its scope of analysis beyond the proposed activity with the Corps jurisdiction...only where the Corps determines that extension to be appropriate under its NEPA regulations and other relevant authorities.[68]

What are unknown are which activities the Corps may determine are beyond its jurisdiction, but appropriate for consideration under NEPA. So far, at least, there appears to be some disagreement between EPA and the Corps over the potential project impacts that should be addressed during the NEPA process.

In an April 2012 letter to the Corps, EPA notes various elements of the Coyote Island terminal project that led EPA to determine that an EIS was necessary for that project.[69] That determination alone conflicts with the Corps' determination that the project would, at least initially, require the preparation of an EA.

Additionally, in delineating the project impacts that warrant the preparation of an EIS, EPA identified the potential public health impacts from coal dust and diesel pollution related to the proposed Coyote Island terminal project; the high level of interest and concern among communities, agencies, interest groups, and industries regarding the proposal to transport coal from Wyoming and Montana to Asia; the uncertainty of potential impacts of transporting large quantities of PRB coal, including trans loading activities on the Columbia River; and the proposed and potential future projects' contribution to cumulatively significant impacts to human health and the environment from increases in greenhouse gas emissions, rail traffic, mining activities on public lands, and the transport of particulate matter and mercury from Asia to the United States, among other possible impacts.

To address the cumulative impacts, EPA recommended that the Corps conduct a broadly scoped cumulative impacts analysis of exporting large quantities of PRB-mined coal through the U.S. West Coast to Asia. The Corps has taken another approach and, as stated earlier, the Corps is preparing an EA for the Coyote Island terminal project and individual EISs for the other pending projects.

EPA's assessment of potential project effects that should be included for analysis appears to be broader than the project-specific effects that the Corps would address in the NEPA documents for the three currently pending projects.

In addition to the April 2012 letter from EPA, some Members of Congress have also written to the Corps requesting that the scope of its environmental review include a comprehensive analysis of the cumulative impacts of all proposed port terminal projects.

The Corps accepted comments from the public until early 2013 regarding the potential scope of issues it will analyze in the GPT and MBT EISs. It may be sometime after the scoping process is complete before it is known how extensively the Corps will consider indirect and cumulative impacts of the proposed projects. In June 2013 the Corps announced that it would not conduct an area-wide EIS of the potential impacts of the three proposed export terminals.

ISSUES AND INTERESTS

The Obama Administration continues to support clean coal, natural gas, renewables, and uranium as energy sources for electric power in the United States, as U.S. coal exports continue to rise. The Administration has not been clear where it falls along the spectrum of coal exports but announced on June 25, 2013, a proposal to limit U.S. government financing of coal plants overseas as part of an overall U.S. strategy to reduce carbon emissions.

As U.S. hydrocarbon resources have expanded over the last few years, there is greater interest by some groups, mainly producers, to export a portion of these resources.

In addition to coal, the United States became a net exporter of petroleum products in 2011, and there are 23 pending liquefied natural gas export projects at various stages of the regulatory approval process. There has even been interest in possibly exporting crude oil from the United States in the future.

Planned expansion of U.S. hydrocarbon exports has generated some controversy. Environmental groups and various consumer groups have been most vocal against exports. Coal exports have attracted greater attention from both sides of the question as fast as new projects have been proposed. In addition to areas where new port capacity is being proposed, some communities along potential and existing rail routes, and mining sites, have voiced their opposition.

CONGRESSIONAL ACTION

Coal Export Legislation

- H.R. 2396, the True Cost of Coal Act of 2013, would extend to 50 years the recovery period, for depreciation purposes, for specified coal port property used for the export of coal.
- S. 831, the Coal Miner Employment and Domestic Energy Infrastructure Protection Act of 2013, would prohibit the Secretary of the Interior, before December 31, 2017, from issuing or approving any proposed or final regulation under the Surface Mining Control and Reclamation Act of 1977 that would: (3) reduce the quantity of coal available for domestic consumption or for export.

Hearings

- House Energy and Commerce Committee, Subcommittee on Energy and Power, "U.S. Energy Abundance Regulatory, Market, and Legal Barriers to Export," June 18, 2013, addressed coal exports.
- Science, Space, and Technology Committee, Subcommittee on Energy, "The Future of Coal: Utilizing America's Abundant Energy Resources," July 25, 2013.

ADDITIONAL READING

U.S. Coal Exports: National and State Economic Contributions, Prepared for the National Mining Association, by Ernst and Young, May 2013.

Our Pain, Their Gain: Mountains Destroyed for Coal Shipped Overseas, by U.S. House Committee on Natural Resources—Democrats, July 19, 2012.

APPENDIX A. MAP OF COAL DEPOSITS AND INFRASTRUCTURE

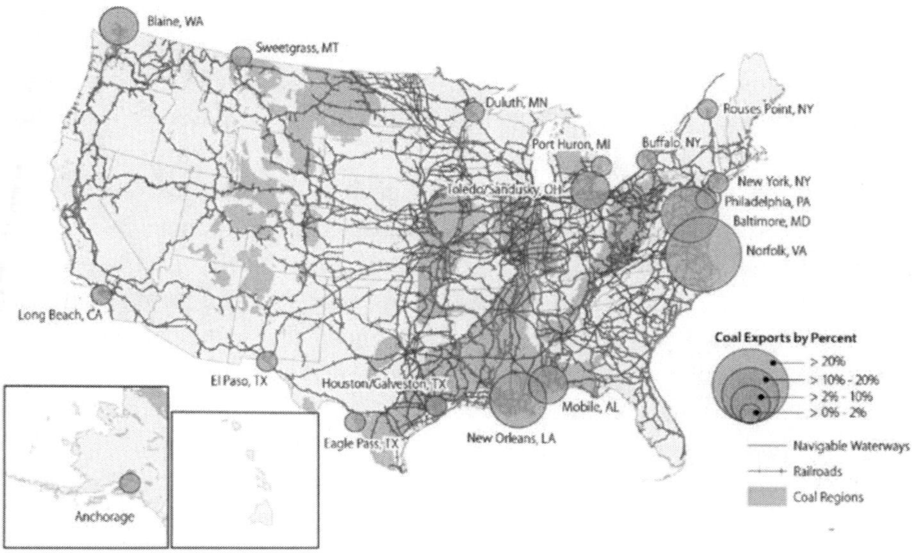

Source: Compiled by CRS.

Figure A-1. Coal Deposits and Infrastructure.

APPENDIX B. CHARACTERISTICS OF DIFFERENT COALS

Coal—a dense carbonaceous fossil fuel—is formed from decayed organic matter that has been subjected to various temperatures and pressures without the presence of oxygen. This burnable rock also contains quantities of hydrogen, sulfur, and nitrogen, and other elements and various amounts of mineral matter. Coal seams are formed along with other sedimentary rocks, primarily sandstone and shale. There are four basic types of coal with varying characteristics throughout the United States. The characteristics of a particular coal will determine how it is used, mainly for heat or electric power generation, or in steel production. Coal quality is measured by its energy value (e.g., British thermal units or Btus per pound),[70] moisture and sulfur levels, and ash content.

Lignite: A brownish-black coal with high moisture and ash content and relatively low heating value, between 4,000 British thermal units (Btu) to 8,300 Btu per pound. It contains the lowest carbon content of the four types of coal, and is usually consumed in electricity generation. Lignite is mainly mined in Texas, North Dakota, Louisiana, and Montana.

Sub-bituminous: This type of coal was the most produced coal in 2011, accounting for 47% of U.S. production. It is a dull black coal with higher heating value than lignite and used for generating electricity and space heat. Its Btu content ranges from 8,300 Btu to 13,000 Btu per pound. Resources are primarily found in Montana, Wyoming, Colorado, New Mexico, Washington, and Alaska.

Bituminous (soft coal): This type of coal is the most abundant in the United States and has a higher heating value than sub-bituminous and lignite, between 10,500 Btu and 15,500 Btu per pound. It is typically used for electric power generation in the United States. Coke is also produced from bituminous coal. Metallurgical coke produced from bituminous coal is used to make steel. It is found primarily in Appalachia and the Midwest.

Anthracite (hard coal): Anthracite has the highest carbon content and energy content (15,000 Btu per pound) of all coals but occurs in limited geographic areas, mainly in Appalachia and Pennsylvania. The highest grades are used in metallurgy.

Thermal Coal: Coal that is used primarily to generate steam or heat for industrial purposes. Used in electric power generation.

Metallurgical Coal: Also known as coking coal, this type of coal is used to make steel.

Appalachia: Primarily the East Coast of the United States, and includes the North, Central, and Southern Appalachia Basins.

Interior: Mainly in the Midwest part of the United States and includes the Illinois Basin.

Western: Mostly the Rocky Mountain region of the United States and includes the Powder River and Uinta Basins.

End Notes

[1] For additional information on global coal production, see CRS Report R43011, *U.S. and World Coal Production, Federal Taxes, and Incentives*, coordinated by Marc Humphries, and for additional information on U.S. coal consumption, see CRS Report R42950, *Prospects for Coal in Electric Power and Industry*, by Richard J. Campbell, Peter Folger, and Phillip Brown.

[2] U.S. Energy Information Administration, *Annual Energy Outlook 2013 Early Release*, DOE/EIA-0383ER(2013), Washington, DC, December 5, 2012, http://www.eia.gov/oiaf

/aeo/tablebrowser/#release=AEO2013ER&subject=7-AEO2013ER&table=15-AEO2013 ER&region=0-0&cases=early2013-d102312a.

[3] International Energy Agency, *Coal Medium-Term Market Report 2012*, Paris, France, December 18, 2012, p. 12.

[4] Laszlo Varro, Head of Gas, Coal and Power Markets, *Medium Term Outlook for Coal presentation*, International Energy Agency, January 24, 2013, p. 3.

[5] The data provided in the report are in short tons (st), metric ton (mt) and in million tons of oil equivalent (Mtoe) because of the different reporting agencies. Tonnes = metric tons.

[6] International Energy Agency, *World Energy Outlook 2011*, Paris, France, November 9, 2011, p. 419, http://www.worldenergyoutlook.org/publications/weo-2011/.

[7] Units differ depending upon the source of information, which is why data differ. In this section, figures are cited in metric tonnes instead of short tons as is U.S. convention. 1 metric ton = 2,205 lbs.

[8] Laszlo Varro, Head of Gas, Coal and Power Markets, IEA, "Medium Term Outlook for Coal," Presentation at CSIS, Washington, DC, January 24, 2013, p. 3.

[9] Primary energy, as defined in the *BP Statistical Review of World Energy, 2012*, comprises commercially traded fuels, including modern renewables used to generate electricity.

[10] *BP Statistical Review of World Energy, 2012*, London, June 2012, p. 41, http://www.bp.com /statisticalreview.

[11] *BP Statistical Review of World Energy, 2012*, http://www.bp.com/sectionbodycopy.do? categoryId=7500&contentId= 7068481, and the U.S. Energy Information Administration, *International Energy Statistics*, accessed September 14, 2012, http://www.eia.gov/cfapps /ipdbproject/IEDIndex3.cfm.

[12] U.S. Energy Information Administration, *International Energy Statistics*, http://www.eia.gov /cfapps/ipdbproject/ IEDIndex3.cfm.

[13] Global imports divided by consumption.

[14] U.S. Energy Information Administration, *International Energy Statistics*, http://www.eia.gov /cfapps/ipdbproject/ iedindex3.cfm?tid=1&pid=1&aid=4&cid=regions,&syid=2006&eyid= 2010&unit=TST.

[15] Laszlo Varro, Head of Gas, Coal and Power Markets, IEA, "Medium Term Outlook for Coal" Presentation at the Center for Strategic and International Studies (CSIS), Washington, DC, January 24, 2013.

[16] Volatiles are defined as hydrogen, methane, carbon monoxide, and other hydrocarbons.

[17] Office of Technology Assessment, *Coal Exports and Port Development*, NTIS order #PB81-203358, April 1981, p. 33.

[18] U.S. Energy Information Administration, *U.S. Coal Exports by Country of Destination, Annual, database*, October 19, 2011, http://www.eia.gov/totalenergy/data/annual/pdf /sec7_13.pdf.

[19] See Appendix B for explanation of thermal and metallurgical coal.

[20] U.S. Energy Information Administration, *Historical Coal Exports database*, May 1, 2012, http://www.eia.gov/coal/ data.cfm#imports.

[21] The United States has three major coal producing regions: Appalachian, Interior, and Western

[22] U.S. Energy Information Administration, http://www.eia.gov/coal/production/quarterly /pdf/t7p01p1.pdf.

[23] Customs districts include a wider geographic area than just the city or town. See http://www.census.gov/foreigntrade/schedules/d/dist.txt.

[24] Brown coal or lignite is a type of coal with low energy content.

[25] IEA, *Coal Information 2011—With 2010 Data*, December 2011.

[26] Ibid. Note: EIA assumes existing statutes and regulations in its reference case at the time it is made.

[27] See Table 15 in EIA, *Annual Energy Outlook 2012*, June 25, 2012, http://www.eia.gov /forecasts/archive/aeo12/ source_coal.cfm.

[28] U.S. Energy Information Administration, *Annual Coal Distribution Report*, Washington, DC, November 30, 2011, http://www.eia.gov/coal/distribution/annual/.

[29] It is unclear how much of the remainder went to export or into domestic stockpiles. U.S. Energy Information Administration, *Quarterly Coal Distribution Report*, Washington, DC, April 16, 2012, http://www.eia.gov/coal/ distribution/quarterly/.

[30] Jeremy Fugleberg, "Utility's cuts could trim demand for Wyoming's Powder River Basin coal," Wyoming Star Tribune, June 21, 2011.

[31] Bruce Kelly, "Which way(s) west for coal?" *Railway Age*, March 2012, p. 18.

[32] "Thermal Coal Terminal Proposed for Fraser River," Vancouver Sun, Nov. 24, 2012, p. H-1.

[33] A permit from the Corps is needed for any project that discharges dredge or fill material in waters of the United States or wetlands, pursuant to provisions in Section 404 of Clean Water Act; and for the construction of any structure in, over, or under navigable waterways of the United States, including excavation, dredging, or deposition of these material in these waters, pursuant to Section 10 of Harbors Act of 1899. The proposed projects in Washington and Oregon will involve such activities and must obtain either or both a Section 404 and Section 10 permit from the Corps, before the project can proceed. Discussion of the Corps permit requirements is beyond the scope of this report. For information regarding applicable authorities, obligations, or regulatory requirements applicable to the Corps obligations to permit coal export terminals, contact Nicole Carter at 7-0854.

[34] Information about the Gateway Pacific Terminal is available at http://gatewaypacific terminal.com/.

[35] Information about Ambre's Millennium Bulk Terminal and Coyote Island Terminal projects are available at http://ambreenergy.com/projects/millennium.

[36] Beth Ward, "Outlook for Western and US Gulf coal," *Platts*, October 5, 2012, "online edition."

[37] "Drought Weighs On Great Lakes Coal Shipments," *Coal Transportation,* Sept. 18, 2012, p. 5

[38] Their exceptional weight puts wear and tear on switches and varying train speed burns more fuel.

[39] SNL Coal Report, *U.S. Coal Export Capacity to Grow Nearly 30% by 2015,* July 16, 2012.

[40] As per 49 U.S.C. 10501(b).

[41] 49 U.S.C. 10901 et seq., 49 U.S.C. 11321 et seq.

[42] 23 U.S.C. 130; 23 C.F.R. 646.210.

[43] Union Pacific and BNSF are the two large (Class I) railroads whose networks cover the western United States.

[44] Testimony of Janssen Thompson, General Manager, Powder River Division, BNSF Railway Co., U.S. Congress, House Committee on Resources, Subcommittee on Water and Power, hearing, *Keeping the Lights On and Maintaining Wyoming's Jobs: Overcoming the Challenges Facing Western Power Generation Facilities*, 109th Cong., 2nd sess., August 9, 2006, p. 2.

[45] Ballast is the bed of crushed stone under the track.

[46] See STB Finance Docket #s FD 35305 and FD 35557; http://www.stb.dot.gov.

[47] See STB Finance Docket # FD 30186. This project dates back to the late 1990s.

[48] For further discussion, see CRS Report R41430, *Inland Waterways: Recent Proposals and Issues for Congress*, by Charles V. Stern.

[49] See CRS Report R41042, *Harbor Maintenance Trust Fund Expenditures*, by John Frittelli.

[50] For further information on the Corps responsibility for waterside infrastructure, see CRS Report R41042, *Harbor Maintenance Trust Fund Expenditures*, by John Frittelli, and CRS Report R41961, *Army Corps Fiscal Challenges: Frequently Asked Questions*, by Nicole T. Carter and Charles V. Stern.

[51] The tax was originally also assessed on exported cargo but this practice was declared unconstitutional.

[52] UNCTAD, *Review of Maritime Transport 2011*, pp. 72-75.

[53] Among the environmental groups are the Sierra Club, Puget Soundkeeper Alliance, Earthjustice, and Columbia Riverkeeper.

[54] In the United States District Court for the Western District of Washington, Case 2:13-cv-00967-JCC, Filed 6/04/13. The plaintiffs argue that the coal trains are discharging coal pollutants into waters throughout Washington.

[55] Concerns common to communities along the coal transport corridors can be seen in statements made by Oregon's Multnomah County Chair, Jeff Cogen in September 10, 2012, remarks to the Multnomah County Health Department requesting a study of the effects of coal train export through the county, available at http://web.multco.us/news/chairjeff-cogens-prepared-remarks-coal-train-export-through-multnomah-county.

[56] For example, releases of coal dust to the air or its deposition on adjacent structures, land, or surface water could be minimized or largely eliminated through the use of enclosed barges, construction of enclosed loading and/or unloading terminal facilities, or the use of enclosed train cars or implementation of certain coal handling practices (the application of surfactants or loading coal to reduce drag on coal piles).

[57] There is virtual unanimity among scientists that greenhouse gases in the atmosphere warm the Earth's climate, though some disagree whether rising GHG concentrations are the largest influence on certain time scales (e.g., of a human lifetime). Numerous scientific organizations have addressed this topic, including the National Research Council, *Advancing the Science of Climate Change*. America's Climate Choices: Panel on Advancing the Science of Climate Change. Washington, D.C.: The National Academies Press, 2010.

[58] See, among other reports, CRS Report R41561, *EPA Regulations: Too Much, Too Little, or On Track?*, by James E. McCarthy and Claudia Copeland.

[59] For more information on the UNFCCC, see CRS Report R40001, *A U.S.-Centric Chronology of the International Climate Change Negotiations*, by Jane A. Leggett. The UNFCCC contains dispute resolution mechanisms but no penalties.

[60] See the Senate's Byrd-Hagel Resolution, S.Res. 98 (July 25, 1997) and the Sense of the Congress on Climate Change of the 108th Congress in Section 1001, H.R. 6, agreed by both chambers and conference committee in 2003.

[61] Coal combustion in China, India, and other countries is also a major cause of acute air pollution, leading to hundreds of thousands of cases annually of morbidity and mortality linked with respiratory, cardiovascular, and cerebrovascular illness and diseases, according to the World Bank (2007) (http://siteresources.worldbank.org/ INTEAPREGTOPENVIRONMENT/Resources/China_Cost_of_Pollution.pdf). The Chinese central government has promulgated stricter standards for emissions and air quality in recent years but struggles to obtain compliance with them.

[62] See http://www.businessweek.com/ap/2012-08-15/india-to-get-9m-tons-of-appalachian-coal-annually.

[63] See, for example, the discussion in Todd Allen, "Alpha Natural Resources Third Quarter 2012 Conference Call" (November 2, 2012) at http://seekingalpha.com/article/974681-alpha-natural-resources-ceo-discusses-q3-2012-resultsearnings-call-transcript?part=single.

[64] Although the emphasis of this section is on possible exports through the Pacific Northwest, the same rules apply to coal export terminals regardless of their location. The proposed projects in the Pacific Northwest, though, would be new facilities, whereas the Gulf Coast and Northeast proposed projects focus on expansions of existing facilities that may require less stringent regulatory approvals.

[65] For information about NEPA requirements, including those regarding preparation of an EIS or EA, see CRS Report RL33152, *The National Environmental Policy Act (NEPA): Background and Implementation*, by Linda Luther.

[66] Corps regulations implementing NEPA are found at 33 C.F.R. §230; actions identified as those normally requiring an EA are listed at §230.7(a).

[67] 40 C.F.R. §1508.25; indirect effects include those at that may occur later in time or farther removed in distance, but are still reasonably foreseeable; cumulative effects include those that result from the incremental impact of the action when added to other past, present, and reasonably foreseeable future actions regardless of what agency (federal or nonfederal) or person undertakes such other actions.

[68] Letter from Jo-Ellen Darcy, Assistant Secretary of the Army, to Governor John A. Kitzhaber, M.D., July 8, 2012, available online.

[69] In framing the proposed project's impacts, EPA structured a significance determination using regulatory criteria, at 40 C.F.R § 1508.27, that are used to identify the context and the intensity of a project's impact.

[70] British thermal unit or Btu is defined as the amount of energy required to heat one pound of water from 39/ F to 40/ F.

In: United States Fossil Fuel Exports
Editor: Mitchell P. Henning

ISBN: 978-1-63117-198-7

Chapter 4

COAL: A KEY PLAYER IN EXPANDED U.S. ENERGY EXPORTS*

Yemi Assefa, Helen McCulley, Myron Murray, and Sharon Royales

INTRODUCTION

In a November 2012 report that received significant attention in the media, the International Energy Agency (IEA) predicted that the United States will be nearly energy self-sufficient by the year 2035.[1] Justifications for this claim include reports of declining oil consumption, the unlocking of natural gas resources through new technologies, and overall increases in U.S. energy production.[2] As natural gas becomes more inexpensive and renewable energy sources continue to increase in usage, the energy mix in the United States (and other countries) is expected to change. The IEA estimates that 40 percent of world electricity generation was fueled by coal in 2011.[3] Coal is the only major energy source for which the United States demonstrates a trade surplus.[4] In 2011, the last calendar year for which data are available, the United States exported more than $16 billion worth of coal, while importing only slightly less than $3 billion in coal. Furthermore, coal is expected to remain one of the largest fuel sources in worldwide energy consumption for at least the next two

---

* This article was published in Beyond The Numbers: Global Economy, (Volume 2/Number 3, February 2013), a publication of the U.S. Bureau of Labor Statistics.

decades. Second only to oil in meeting the energy needs of the world, coal generates more electricity for the United States and the world than any other single fuel. With the key role that coal currently plays in the U.S. trade balance, and is expected to play in the future, it is interesting to look at recent volatile export price movements of this important U.S. resource.

The Recent Growth in U.S. Coal Exports

Although most of the coal produced in the United States is consumed domestically, export volumes of coal have increased in recent years. The proportion of coal production going toward exports has also increased, doubling from 5 percent in 2009 to 10 percent in 2011. The majority of coal production consists of thermal or steam coal that goes to electricity generation, but metallurgical or coking coal is primarily used in making iron and steel. In fact, nearly 70 percent of global steel production depends on this grade of coal.[5] Consequently, in 2011, metallurgical coal exports accounted for 77 percent of U.S. coal exports in terms of trade dollars and 65 percent in terms of volume.[6] As seen in chart 1, export coal prices began to climb in early 2010. After rising only 5.9 percent over the previous 24 months, the BLS coal and related fuels export price index rose 51.8 percent during the first 6 months of 2010. The index for metallurgical coal increased 71.3 percent over this period. Prices for both coal and related fuels and metallurgical coal continued to advance at a more modest pace until December 2010. Overall for 2010, the indexes increased 51.6 and 68.2 percent, respectively.

U.S. coal producers were able to increase both export volumes and prices in 2010 because of growing energy needs in markets such as China and India and the reduced world coal supply caused by heavy rains and flooding in major coal exporting countries.[7] In 2010, exports to Asian markets increased 176 percent from 2009 levels, primarily because of a surge in exports of metallurgical coal to China, Japan, and South Korea.[8] Metallurgical coal exports accounted for 83 percent of the growth in 2010 export volumes.

Export prices for coal continued to increase until September 2011. In August 2011, the year-to-date increase in export prices stood at 23.7 percent. Prices then declined 7.0 percent over the remaining months of the year, ending with a 15.1-percent increase overall for 2011. Metallurgical coal prices recorded a similar trend, increasing 27.1 percent for the first 8 months in 2011, before declining 7.3 percent for the remainder of the year, and finishing with a 17.8-percent advance for the year.

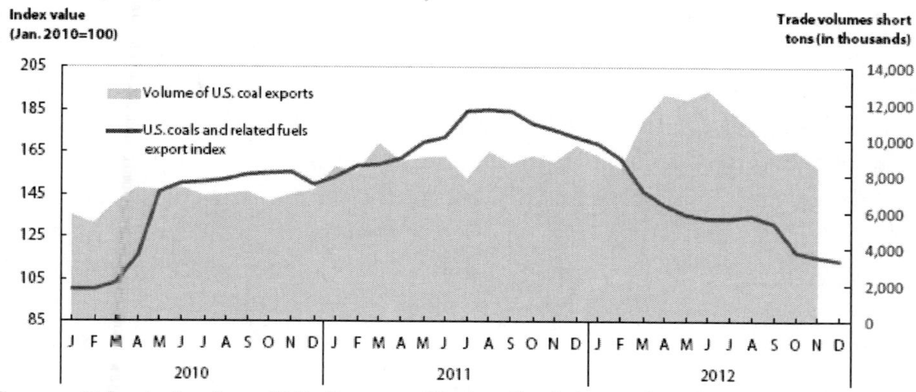

Source: Price index from U.S. Bureau of Labor Statistics and trade volumes from the Energy Information Administration Monthly Energy Review .

Chart 1. U S. coal exports: price index and volumes, January 2010–December 2012.

The volume of U.S. coal exports was 31.3 percent higher in 2011 than in 2010.[9] The expansion of coal-fired power plants in both China and India contributed to the increase in international demand for coal. However, U.S. domestic demand was limited by competition from natural gas. With higher sales prices in Asia and Europe in 2011, U.S. coal exporters diverted the highest proportion of coal production to exports since 1992.[10]

As seen in chart 2, the proportion of coal trade dollars coming from Asia experienced sizable increases in 2010 and 2011. The dollar value of U.S. exports to only China and India increased by more than $1.8 billion from 2002 to 2011.[11] Overall, the export value of coal exports increased by almost $15 billion over this same period.[12]

A Reversal in Coal Prices

In contrast, in 2012, export coal prices turned down sharply. The export index fell 33.9 percent from December 2011 to December 2012, after surging 74.4 percent for the 2 preceding years. The declining export price of coal was attributable to such factors as decreased domestic demand, a slowdown in economic growth in both China and India, and the continuing weakness seen in European economies. Although U.S. coal production fell in 2012, export volumes grew to record levels. Even as the United States transitions away from coal usage, certain European nations are relying more heavily on coal. With oil and natural gas prices still relatively strong in Europe, inexpensive

U.S. coal is more in demand by European utility companies.[13] The price for carbon emissions permits in the European Union has also fallen, making coal even more attractive to utilities. Compared with 2011, exports to Europe were up 29.4 percent for the year by the third quarter of 2012, with the United Kingdom and Italy accounting for more than half of the increase.[14] Exports to Asia were up 23.7 percent for the year by the third quarter of 2012, mainly because China doubled the amount of coal imported from the United States.[15]

According to the U.S. Energy Information Administration, in 2013, U.S. export coal volumes are expected to remain above 100 million short tons for the third straight year and international coal prices are expected to be lower.[16] In the coming year, U.S. export coal prices will be influenced by the difference between domestic and international natural gas prices, increased coal use in newly industrialized economies, and Asian domestic production levels for coal, along with other factors. BLS will continue tracking export prices for this historically important U.S. commodity in order to identify market and economic trends.

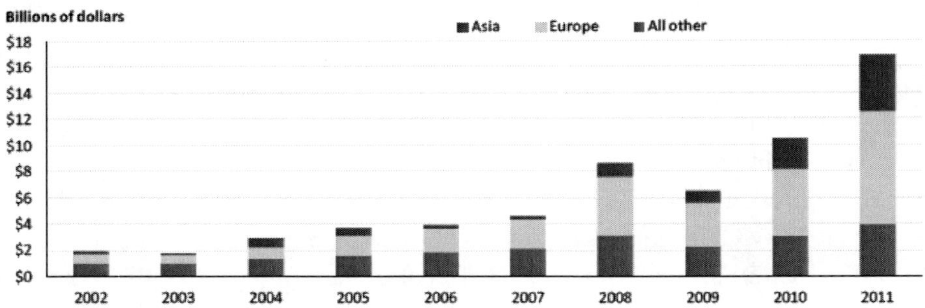

Source: U.S. Bureau of Labor Statistics.

Chart 2. U.S. export coal trade dollar values in billions of dollars for Asia, Europe, and all other regions, 2002–2011.

FOURTH QUARTER 2012 HIGHLIGHTS

Import Prices

The price index for overall imports declined 1.5 percent in 2012, following an increase of 8.5 percent for the previous year and a 5.3-percent advance in 2010. (See table 1.) The decrease in import prices was driven by a

6.4-percent decline in the price index for fuel imports. This was the first calendar-year decline in import prices since 2008.

Fuel import prices. Prices for fuel imports (which account for nearly one-quarter of the overall import prices index) fell 6.4 percent in 2012, the first calendar-year decrease since a 47.0-percent drop in 2008. During the final quarter of 2012, the price index for fuel imports declined 2.8 percent.

Petroleum prices led the overall decline in fuel prices in 2012, falling 7.1 percent for the year. Following advances of 27.1 percent and 16.2 percent for the 2 previous years, prices for import petroleum trended mostly down throughout 2012, declining 4.1 percent in the final quarter. Both supply concerns and weak demand contributed to the lower prices. Concerns about the U.S. and European economies contributed to weak worldwide demand. In particular, worries remained about the direction of the U.S. economy (the world's largest) and the so-called fiscal cliff, until a compromise was reached by the U.S. government at the beginning of 2013.

In addition, despite a pledge to rein in overproduction, the Organization of the Petroleum Exporting Countries (OPEC) produced more crude oil than forecasted for demand in 2012.[17] This took place in spite of growing competition from both the United States—which reached the highest level of oil production in nearly 15 years—and from Russia.[18] In fact, the U.S. Energy Information Administration predicted that net imports of liquid fuels, including crude oil and petroleum products, would fall to their lowest level since 1987 and would post numbers that were only half of the peak levels reached between 2004 and 2007.

For the year, natural gas prices advanced 14.2 percent, the first calendar-year increase since 2005, when the index rose 54.9 percent. The price index for natural gas rose 50.3 percent during the final quarter of 2012, after increasing 18.4 percent and 18.0 percent in the last 2 months of the year. Cooler-than-average temperatures during the end of 2012 produced strong demand and pushed up overall prices.[19]

Nonfuel import prices. In contrast to fuel prices, nonfuel prices ticked up 0.1 percent in 2012, the smallest yearend increase since the index recorded no change in 2002. As shown in chart 3, higher prices for automotive vehicles were the most significant contributor to the overall increase in nonfuel import prices. During 2012, the price index for capital goods remained unchanged, while consumer goods prices edged down 0.1 percent and prices for foods, feeds, and beverages fell 1.7 percent.

Prices for automotive vehicles advanced 1.9 percent in 2012, following a 3.3-percent increase the previous year. The 2012 increase was driven by a 1.3-

percent rise in the first 4 months of the year. Higher prices for passenger cars, up 2.3 percent, led the overall increase in automotive vehicles prices.

Table 1. Summary of December-to-December percent changes in all-commodity imports and exports, fuel and agricultural indexes, 2003–2012

Index	2003	2004	2005	2006	2007	2008	2009	2010	2011	2012
All-commodity imports	2.4	6.7	8	2.5	10.6	-10.1	8.6	5.3	8.5	-1.5
Excluding fuels	1	3	1.1	2.9	3.1	1.2	0.3	3	3.4	0.1
Fuels	13.2	31.5	43.5	0.9	42.1	-47	62.2	14.2	24.9	-6.4
All-commodity exports	2.2	4	2.8	4.5	6	-2.9	3.4	6.5	3.6	1.1
Excluding agricultural	1.3	5	2.6	3.7	4.5	-2.2	2.9	5.1	4	-0.4
Agricultural	13.4	-5.9	4.9	13.5	23.3	-10.9	9.2	20.5	1	12.8

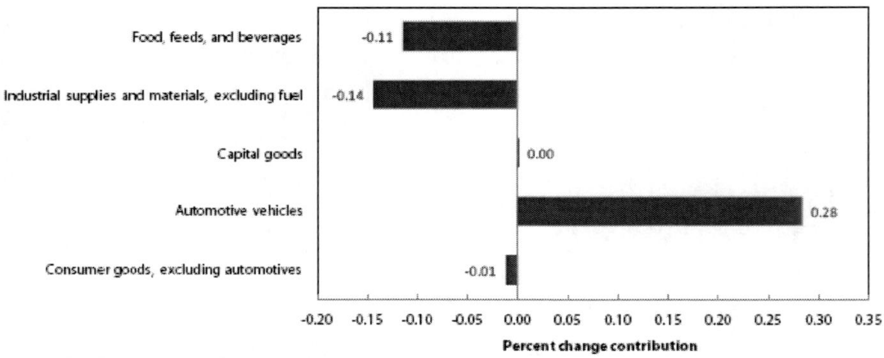

Source: U.S. Bureau of Labor Statistics.
Note: Due to rounding, figures do not add to total.

Chart 3. Major contributors to the 0.1-percent increase in import prices, excluding fuel, 2012.

The price index for nonfuel industrial supplies and materials decreased 0.8 percent in 2012, the first calendar-year decline since a 0.2-percent drop in 2008. The decline in 2012 followed a 6.7-percent increase in 2011 and a 12.2-percent advance in 2010. A 4.9-percent decrease in unfinished metals prices led to the overall decline.

Prices for foods, feeds, and beverages fell 1.7 percent in 2012, after rising 6.0 percent the previous year. The 2012 decrease was the largest calendar-year drop since a 4.7-percent decline in 2001. The price index for coffee led the

2012 decrease, falling 26.8 percent over the year. As the 2012–2013 crop season started in October, the International Coffee Organization reported increased production in several exporting countries, especially Brazil, which drove prices down.[20]

Export Prices

U.S. export prices rose 1.1 percent in 2012, the smallest calendar-year increase since a 2.9-percent decrease in 2008. The increase in 2012 followed a 3.6-percent increase in 2011 and a 6.5-percent advance in 2010. A 12.8-percent jump in the price index for agricultural commodities more than offset a 0.4-percent decline in nonagricultural commodities prices.

Agricultural export prices. Prices for agricultural exports increased 12.8 percent in 2012, following a 1.0-percent advance in 2011 and a rise of 20.5 percent in 2010. Over the last 4 years, the index has increased by more than 49 percent. Higher prices for soybeans, corn, and wheat, up 31.6 percent, 26.9 percent, and 28.0 percent, respectively, all contributed to the overall increase in agricultural export prices. The drought that hit most of the United States in 2012 severely affected agricultural crops across the nation, with most of the price increases occurring during the summer. As a result of drought-related crop damage, U.S. export prices for corn soared nearly 128 percent above the 20-year historical average.[21] Meanwhile, droughts in Russia, Ukraine, and Kazakhstan cut global wheat stocks for the 2012–2013 season.[22]

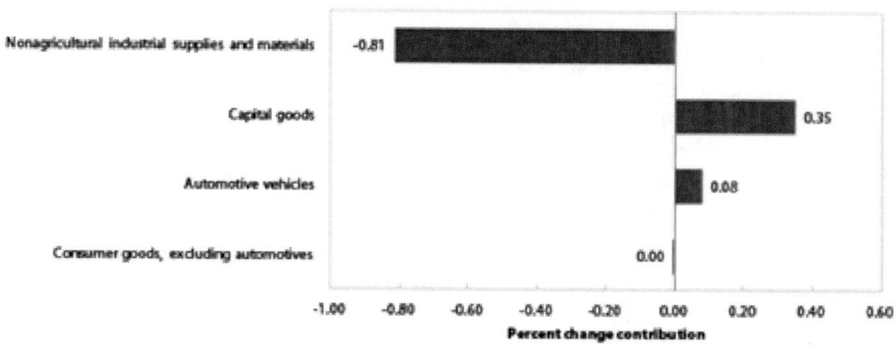

Source: U.S. Bureau of Labor Statistics.
Note: Due to rounding, figures do not add to total.

Chart 4. Major contributors to the 0.4-percent decrease in export prices, excluding agriculture, 2012.

Nonagricultural export prices. The price index for nonagricultural exports declined 0.4 percent over the past 12 months, following a 4.0-percent increase in 2011. As seen in chart 4, a 2.2-percent drop in nonagricultural industrial supplies and materials drove the decrease. This drop in prices for nonagricultural industrial supplies and materials more than offset rising prices for capital goods and automotive vehicles.

The 2012 decrease in nonagricultural industrial supplies and materials prices followed an 8.0-percent advance in 2011 and was the largest calendar-year decrease since a 9.3-percent drop in 2008. Lower prices for fuels and lubricants were the main drivers for a drop in the nonagricultural export price index throughout the year as well as the fourth quarter.

Capital goods prices rose 1.0 percent in 2012. Rising prices for electrical generating equipment were the major factors in the increase. In 2012, automotive vehicles prices advanced 0.9 percent, while prices for consumer goods were unchanged.

End Notes

[1] "World Energy Outlook 2012," International Energy Agency, (November 12, 2012), http://www.worldenergyoutlook.org/publications/weo-2012/.

[2] Donald A. Norman, "U.S. Oil Consumption Declines," Manufacturers Alliance for Productivity and Innovation, (January 8, 2013), http://www.mapi.net/us-oil-consumption-declines.

[3] "Key World Energy Statistics 2012," International Energy Agency, (2012), http://www.iea.org/publications/freepublications/publication/kwes.pdf.

[4] Brock R. Williams and J. Michael Donnelly, "U.S. International Trade: Trends and Forecasts," (Congressional Research Service, October 19, 2012), p. 31, http://www.fas.org/sgp/crs/misc/RL33577.pdf.

[5] "Coal & Steel Facts 2011," World Coal Association, (August 2011), http://www.worldcoal.org/bin/pdf/original_pdf_file/coal_and_steel_facts_2011(17_08_2011).pdf.

[6] "U.S. Coal Exports by Year and Quarter (2002–2011)," (U.S. Energy Information Administration, May 1, 2012), http://www.eia.gov/coal/archive/coal_historical_exports.xls.

[7] William Watson, Nicholas Paduano, Tejasvi Raghuveer, and Sundar Thapa, "U.S. Coal Supply and Demand: 2010 Year in Review," (U.S. Energy Information Administration, June 1, 2011), http://www.eia.gov/coal/review/pdf/feature10.pdf.

[8] Ibid.

[9] Annual Energy Review, (U.S. Energy Information Administration, September 2012), table 7.1, http://www.eia.gov/totalenergy/data/annual/index.cfm#coal.

[10] Monthly Energy Review, (U.S. Energy Information Administration, December 2012), table 6.1, http://www.eia.gov/totalenergy/ data/monthly/pdf/sec6_3.pdf.

[11] U.S. Census Bureau, Microsoft Excel file, "Foreign Trade, Country and Product Trade Data, Product Detail and Partner Country, End-Use, Exports," http://www.census.gov/foreign-trade/statistics/product/enduse/exports/enduse_exports.xls.

[12] Ibid.

[13] Ed Crooks and Sylvia Pfeifer, "US coal exports to Europe soar," The Financial Times, October 3, 2012, http://www.ft.com/intl/cms/s/0/fbf0b9fa-0d63-11e2-97a1-00144feabdc0.html#axzz2IG3q8CCc.

[14] Quarterly Coal Report, DOE/EIA-0121 (2012/03Q) Table 7, (U.S. Energy Information Administration, December 20, 2012).

[15] Ibid.

[16] Short-Term Energy Outlook, (U.S. Energy Information Administration, January 2013), http://www.eia.gov/forecasts/steo/pdf/steo_full.pdf.

[17] Benoit Faucon, "Oil Prices May Fall, Despite Mideast," The Wall Street Journal, November 21, 2012, http://online.wsj.com/article/SB10001424127887324352004578132902620727298.html?KEYWORDS=petroleum+prices.

[18] Tennille Tracy, "U.S. Oil Output Hits Nearly 15-Year High," The Wall Street Journal, December 4, 2012, http://online.wsj.com/article/SB10001424127887323717004578159624213068966.html?KEYWORDS=petroleum+prices.

[19] David Bird, "Natural Gas Continues to Rally," The Wall Street Journal, November 14, 2012, http://online.wsj.com/article/SB10001424127887324735104578118880267975420.html?KEYWORDS=natural+gas+prices+hurricane+sandy.

[20] "Monthly Coffee Market Report, December 2012," International Coffee Organization, (December 2012), http://dev.ico.org/documents/cy2012-13/cmr-1212-e.pdf.

[21] Will Adonizio, Nancy Kook, and Sharon Royales, International Price Program, "Impact of the drought on corn exports: paying the price," Beyond the Number: Global Economy, vol. 1, no. 17 (U.S. Bureau of Labor Statistics, November 2012), http://www.bls.gov/opub/btn/volume-1/impact-of-the-drought-on-corn-exports-paying-the-price.htm.

[22] "Russia drought cuts wheat production," AG Professional, August 8, 2012, http://www.agprofessional.com/resource-centers/wheat/news/Russia-drought-cuts-wheat-production-165442056.html.

INDEX

H

I

J

K

L

Q

R

S

T

U

V

W